T0205388

Green Energy and Technology

Climate change, environmental impact and the limited natural resources urge scientific research and novel technical solutions. The monograph series Green Energy and Technology serves as a publishing platform for scientific and technological approaches to "green"—i.e. environmentally friendly and sustainable—technologies. While a focus lies on energy and power supply, it also covers "green" solutions in industrial engineering and engineering design. Green Energy and Technology addresses researchers, advanced students, technical consultants as well as decision makers in industries and politics. Hence, the level of presentation spans from instructional to highly technical.

More information about this series at http://www.springer.com/series/8059

Hoy-Yen Chan • Kamaruzzaman Sopian
Editors

Renewable Energy in Developing Countries

Local Development and Techno-Economic Aspects

 Springer

Editors
Hoy-Yen Chan
Universiti Kebangsaan Malaysia
Solar Energy Research Institute
Bangi, Selangor, Malaysia

Kamaruzzaman Sopian
Universiti Kebangsaan Malaysia
Solar Energy Research Institute
Bangi, Selangor, Malaysia

ISSN 1865-3529   ISSN 1865-3537   (electronic)
Green Energy and Technology
ISBN 978-3-030-07864-5   ISBN 978-3-319-89809-4   (eBook)
https://doi.org/10.1007/978-3-319-89809-4

This Springer imprint is published by the registered company Springer Nature Switzerland AG.
The registered company address is: Gewerbestrasse 11, 6330 Cham, Switzerland

Preface

The need for a transition towards cleaner and low-carbon energy has strengthened the roles of renewable energy in sustainable energy technologies. However, in terms of technology development, it is important to understand the resources availability, the possibilities, and the limitations of the local context. Technological deployment of renewable energy needs policy, skilled personnel, technology development, and financial resources to be in place, and these could be a challenge for developing countries.

This book intends to share the knowledge on local development, which includes reviewing the status and policy toward sustainable energy and sharing the potentials and barriers through case studies and research from Malaysia, Vietnam, Bangladesh, and Brunei. Part I of the book consists of four chapters, which focus on policy and strategy, while another six chapters in Part II focus on the technology development and feasibility.

In Chap. 1, authors from the ASEAN Centre for Energy give an overview of ASEAN renewable energy policy and development and provide barriers and recommendations based on their analysis. Chapter 2 reviews and assesses the energy sector in Malaysia in terms of accessibility, affordability, and sustainability in energy supply and consumption and concludes with highlights and recommendations. Chapter 3 discusses the roles of renewable energy in achieving sustainable development goals and greenhouse gas emissions target and the challenges of Vietnam in implementing the strategies and policies. On the other hand, being a natural gas exporter, Brunei shares the motivations and experiences in gas saving and emission avoidance from the solar power plant in Chap. 4. Then, Chap. 5 provides a survey report on the perspectives from the indigenous community on solar electrification – the acceptance level and the needs of such modern energy service. Chapters 6 and 7 discuss how Bangladesh could maximize the cultivatable land use for photovoltaic panel installation, and how this technology can benefit the agriculture and livestock sectors in a feasible manner. Besides photovoltaic that generates electricity, another type of solar energy technology is solar thermal, which generates heat. Chapter 8 is a case study in Malaysia that shares the success story on solar hot water system in a large scale for a hospital. Last but not least, Chaps. 9 and 10 are the research studies on biogas production and purification in electricity generation, in which the feasibility studies were based on a local context of Vietnam.

The authors of this book are from the government agencies, universities, and industries that share their perspectives, experiences, and research findings on policy formulation and technology deployment. It is thus hoped that this book will benefit a wide range of communities working in the field of renewable energy in developing countries. Nonetheless, we have learned that the most common and popular renewable energy technology in Asian developing countries is the solar photovoltaic. Not only might this technology be relatively simpler to be applied but could be also due to the fact that the data on local solar irradiance is easier to be obtained, compared to the availabilities of other renewable resources. The potentials of other kind of renewable energy resources need to be explored and so diversifying the energy sources and transiting the energy system toward a low-carbon future.

Finally, we would like to express our sincere gratitude to all the authors for devoting time and effort to this book. Without their passions, motivations, and determinations, publishing a book by gathering authors from different countries and background would never be possible. Also, we would like to thank the editorial team of Springer Nature in assisting us throughout the publication process.

Acknowledgment

Almost all the corresponding authors of this book are the alumni of Asian School of Renewable Energy – a regular training course conducted and hosted by the Solar Energy Research Institute (SERI), National University of Malaysia, since 2007. The training course not only has been transferring the knowledge but also has built up a platform for policy makers, engineers, practitioners, academics, and researchers from developing countries in Asia region to meet up, which has established this collaborative network. Therefore, on behalf of the authors, we would like to thank all the cofounders – the Islamic Educational, Scientific and Cultural Organization (ISESCO); United Nations Educational, Scientific and Cultural Organization (UNESCO); Commission on Science and Technology for Sustainable Development in the South (COMSATS); and the Turkish Cooperation and Coordination Agency (TİKA) – for supporting the course all these years.

Bangi, Malaysia   Hoy-Yen Chan
                                                                                         Kamaruzzaman Sopian

Contents

Part I
Policy and Strategy

Status on Renewable Energy Policy and Development in ASEAN

Aloysius Damar Pranadi, Beni Suryadi, and Badariah Yosiyana

1 Introduction

As mandated up to 2030, on 25 September 2015, a set of goals, namely, Sustainable Development Goals (SDGs), were adopted by countries over the world to end the poverty, protect the planet and ensure the prosperity for all. One of the highlighted goals is focused to ensure the access to affordable, reliable, sustainable and modern energy for all, as countries realised that energy is a central and the most essential solution to nearly every major challenge and opportunity the world faces today (United Nations 2017). In line with the spirit of that goal, in September 2015, ASEAN Member States (AMS)[1] launched a fourth series of regional energy blueprint, entitled as ASEAN Plan of Action for Energy Cooperation (APAEC) 2016–2025 Phase 1: 2016–2020 in Malaysia. This document raised up the backdrop theme of *enhancing energy connectivity and market integration to achieve the energy security, accessibility, affordability and sustainability for all* since its endorsement at 32nd ASEAN Ministers on Energy Meeting (AMEM) held on 23 September 2014 in Vientiane, Lao PDR. Through the APAEC, all the AMS aim to support the implementation of multilateral energy cooperation to advance regional integration and connectivity goals in ASEAN, as they acknowledged that energy is key to the realisation of the ASEAN Economic Community (AEC) which calls for a well-connected ASEAN to drive an integrated, competitive and resilient region (ACE 2015b).

Under the APAEC vision on energy connectivity, ASEAN established seven programme areas (PA), namely, PA 1, ASEAN power grid (APG); PA 2, Trans-

---

[1]An official term for all the countries which located in the Southeast Asia region and joined as a member in the Association of South East Asian Nations (ASEAN).

A. D. Pranadi (✉) · B. Suryadi · B. Yosiyana
ASEAN Centre for Energy (ACE) Building, Jakarta, Indonesia
e-mail: damarpranadi@aseanenergy.org

© Springer International Publishing AG, part of Springer Nature 2018
H.-Y. Chan, K. Sopian (eds.), *Renewable Energy in Developing Countries*,
Green Energy and Technology, https://doi.org/10.1007/978-3-319-89809-4_1

ASEAN Gas Pipeline (TAGP); PA 3, clean coal technology (CCT); PA 4, energy efficiency and conservation (EE&C); PA 5, renewable energy (RE); PA 6, regional energy policy and planning (REPP) and PA 7, civilian nuclear energy (CNE). To address the challenges of sustainable energy growth and climate change, the AMS has been following a deliberate policy of diversifying and using indigenous energy sources efficiently at the national level. To this end, the AMS has developed and implemented several renewable energy initiatives, such as biofuels, solar PV programmes, as well as promoting open trade, facilitation and cooperation in the renewable energy sector (ACE 2015a, b). All of these initiatives are covered under PA 5. renewable energy (RE), where in this area, ASEAN has also already set its regional aspirational target of 23% renewable energy on their total primary energy supply (TPES) by 2025.

Against that target, the ASEAN policy framework on RE is one fundamental pillar in renewable energy development. Through this chapter, an analysis for renewable energy policy recommendation in this region will be conducted through a review of the existing ACE studies, namely, *Renewable Energy Policies in ASEAN, Renewable Energy Development 2006–2014 and Renewable Energy Outlooks for ASEAN*. To ensure the up-to-date data, some developments on RE policies and installed capacity/generation will be covered by the data taken from *the* 5th *ASEAN Energy Outlook* (AEO5) published in 35th AMEM, September 2017. This chapter is to provide policy recommendations through the review exercises that based on the ACE database. Furthermore, an analysis from using the REmap tool was used in recommendation for the region in achieving the RE target 23% by 2025.

2   Energy and Renewable Energy Status in ASEAN

ASEAN, a region with abundant renewable energy, recorded a solar irradiation in the range of 3.6–5.3 kWh/m^2/day, wind speed at range of 1.2 m/s to 6.5 m/s (at certain elevation it could be above 6.5 m/s), geothermal potential at slightly over than 30 GW and huge amount of hydropower and biomass. In addition, ocean energy which does not exist commercially in this region may reach up to a fifth of terawatt in terms of potential. Indonesia, the Philippines and Vietnam have the biggest geothermal potential with 28.9 GW, 1.2 GW and 0.34 GW, respectively. Indonesia's hydropower potential is predicted to be around 75 GW. Indonesia registers the biggest potential of biomass with 32.6 GW, followed by Thailand with 2.5 GW. In addition, except for Brunei Darussalam and Singapore, the remaining AMS has also a big potential of biomass. All the potentials are presented in Fig. 1.

In the last two decades, ASEAN total primary energy supply (TPES) has been rising rapidly from only 238 million tonnes of oil equivalent (Mtoe) in 1990 to 627 Mtoe in 2015. The role of renewable energy in TPES has been significantly increasing from 18 Mtoe in 2000, up to 85 Mtoe in 2015. In terms of share, renewable energy in TPES changed from only 4.7% in 2000 to 13.6% in 2015 of

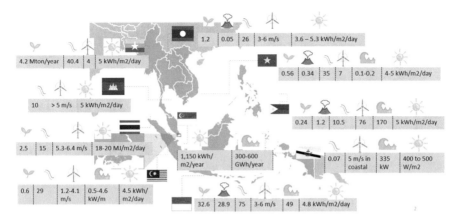

Fig. 1 Renewable energy potential in ASEAN (in GW, else stated). (Source: ACE 2015a, b)

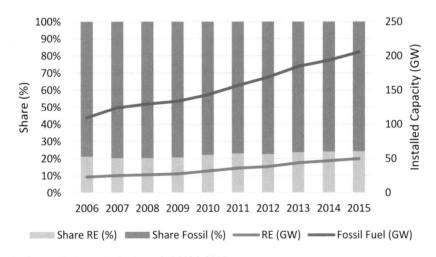

Fig. 2 Installed capacity in the period 2006–2015

which hydro contributed at 18 Mtoe (2.9%), geothermal at 12 Mtoe (1.9%) and other renewable energy at 55 Mtoe (8.8%).

From the 13.6% of RE, power sector is accounted for 7% and rapidly increasing. In 2006, total power installed capacity in ASEAN was only amounting to 109 GW where non-renewable fuels dominated all the share at 79%. In 2015, 205 GW installed capacity was reached all across the region resulting in 24.3% renewable energy share (see Fig. 2).

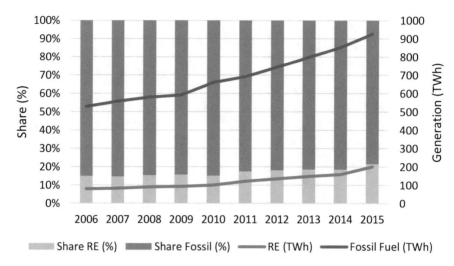

Fig. 3 Power generation in period 2006–2015

In power generation, the renewable energy share had increased to 21.7% of total power generation in 2015, from only 16.2% in 2006. The total power generation in 2006 was only at 529 TWh, while in 2015 it reached at 927 TWh (see Fig. 3). In 2015, renewable energy only contributed around 201 TWh consisting of 146 TWh from hydropower, 27 TWh from biomass, 21 TWh from geothermal, 4 TWh from wind and 3 TWh from solar PV. In 2006, hydropower was only reaching around 61 TWh, biomass merely at 1 TWh, geothermal generated only 17 TWh, solar only 0.002 TWh and wind at 0.005 TWh. Figure 3 presents the growth of power generation in the period 2006–2015.

3 Renewable Energy Policy Development

In ASEAN, the levels of renewable energy policy are broadly varied from one country to other countries. Some countries had successfully established the market for renewable energy, while others are still in the early stage of development in renewable energy policy. *Referring to the Renewable Energy Policies in ASEAN (ACE 2016b)*, the renewable energy policy was categorised into six key policy factors (Table 1), namely, (1) the renewable energy (RE) target, (2) the selling tariff for renewable energy, (3) incentives on power generated from the renewable energy technologies, (4) the financing support for the renewable energy developers, (5) permit and licencing structure for RE power generation and (6) other technical aspects/ standardisation for RE power project connection, such as grid codes. By using these categories, the new updated policies were also classified into six categories of policy with an additional information (see Table 2).

Table 1 Status for renewable energy policy in ASEAN

Policy	Brunei Darussalam	Cambodia	Indonesia	Lao PDR	Malaysia	Myanmar	The Phillipines	Singapore	Thailand	Vietnam
RE target	Yes	Yes	Yes	Yes	Yes	Yes	Yes	Yes	Yes	Yes
Selling tariffs	No	No	Yes	No	Yes	No	Yes	Yes	Yes	Yes
Incentives	No	No	Yes	Yes	Yes	No	Yes	Yes	Yes	Yes
Financing support	No	Yes	Yes	Yes	Yes	No	Yes	Yes	Yes	Yes
Permits and Licences	No	Yes	Yes	Yes	Yes	No	Yes	Yes	Yes	Yes
Technical aspects	No	Yes	Yes	Yes	Yes	No	Yes	Yes	Yes	Yes

Source: ACE (2016a, b)

Table 2 Renewable energy policy in ASEAN, as of 2017 (ASEAN Centre for Energy 2017; Yosiyana (2017))

RE policy by category	Brunei Darussalam	Cambodia	Indonesia	Lao PDR	Malaysia
RE target	10% RE in power generation by 2035	2.24 GW of hydropower by 2020	23% RE share of TPES by 2025	30% RE share of total energy consumption by 2025	2.08 GW by 2020
RE development status	0.03% RE in power generation by 2015	0.93 GW of hydropower by 2015	7.49% RE share of TPES by 2015	23.47% RE share of TFEC by 2015 (include hydropower)	1.2 GW by 2015, exclude hydropower
Selling tariffs	• No feed-in tariffs • Next plan: Feed in tariffs for RE-based power projects and renewable energy certificates (1 MWh above)	• No selling tariffs for renewable energy on-grid system • Only for off-grid system where the tariffs are determined between project developers and consumers (for SHS in rent-to-own basis)	• Tariffs for solar PV, hydro, wind, geothermal and bioenergy (biogas, waste and biomass) – based on electricity production cost (BPP) • Vary on RE types and geographical locations (determined by PLN) • BPP will be updated annually by PLN (utility)	• No feed-in tariff • Electricity tariffs from renewable energy determined between producers and power utility on the case-by-case basis	• Feed-in tariff for bio-mass, biogas, small hydro, solar PV, waste and geothermal • Vary based on RE technology only • Updated annually by sustainable energy development authority (SEDA)
Incentives	No incentives	• 2003 law pronounced several fiscal and investment incentives; however, the way and amount project developers can avail is not	• Tax and custom facility for RE resources utilisation: exemption and deduction for income tax, value added tax, custom and	• Investment promotion law 2009: import duty free on production machinery, equipment and raw materials; import duty free on	• Corporate investment tax allowance and income tax exemption (initial allowance, annual allowance, balancing allowance,

		clear yet • Custom duties are reduced on RE equipment	tax burdened by the government • Access domestic financing for renewable energy development • Financial guarantee on RE projects in case PLN payment failed	chemical materials necessary for biofuels production within 7 years; profit tax is divided in to three categories: 20%, 15% and 10%. Profit tax exemption is possible for a certain period depending on activities, investment areas and size investment; subsidies on unit product price depending on energy type and times period • Hydropower incentives are special in Lao PDR, with some additional fiscal and nonfiscal incentives	balancing charge) • Solar systems are special with additional benefits in terms of tax incentives • Green technology incentive where this incentive allows 100% of qualifying capital • Expenditure incurred on a green technology project to be claimed against 70% of the green technology project income. Unutilised allowances can be carried forward until they are utilised
Financing supports	No financing support	• A subsidy of US$ 100 per systems for rural RE systems. In 2014, a total of US$6 million financed for rural electrification programme • Grant assistance of US$ 400/kWh for any mini/ microhydropower and US$ 300/kWh for other renewables	• Geothermal fund facility provides guarantee funds and loans • Clean technology fund of US$ 400 million and US$ 50 million for geothermal development in 2014 and 2015	Domestic and foreign RE developers may have access to loans from domestic banks and financial institutions, as well as overseas	Green technology financing scheme (GTFS) where 2% loan interest rebate for up to RM50 million of the project loan and 60% guarantee of the financing

(continued)

Table 2 (continued)

RE policy by category	Brunei Darussalam	Cambodia	Indonesia	Lao PDR	Malaysia
Permits and licences	No permits and licences	Electricity law of 2001 provides guidelines for issuing power licences and capacity licences (technical, safety and environment standard)	• Licences for geothermal activities under geothermal law • In 2015, permit procedures of electricity power plant into integrated one-stop service stationed in the investment coordinating board • Electricity power supply business area also was defined under RUPTL 2017–2026 – the updated one	• Licencing approval in RE enterprises	Permits under renewable energy rules 2011 and 2013
Technical standards	No technical standard applied	Grid code 2009 but no specific provisions for renewable energy	Guidelines for connecting renewable energy generation plants to PLN's distribution systems, 2014	Technical Standards for Hydro Development	• Technical standards under renewable energy rules 2011 and 2013 • Feed-in approval and requirements • Electricity generation licences and permits from electricity regulatory commission

RE policy by category	Myanmar	Philippines	Singapore	Thailand	Vietnam
RE target	38% of hydropower and 9% of RE in energy mix by 2030–2031	15.2 GW RE installed capacity in 2030	350 MWp solar installation by 2020 and 10,140 ton/day by 2018 for waste to energy (WtE)	30% RE in total energy consumption by 2036	21% RE of 130 GW installed capacity in 2030
RE development status	6.36% of hydropower and less than 0.1% of RE in energy mix by 2014	6.96 GW of RE installed capacity in 2016	99.9 MWp solar installation and 256.8 MW waste to energy by 2016	11.67% RE in TFEC by 2015	12.22% RE of 130 GW installed capacity by 2015
Selling tariffs	• No selling tariffs from government • Fixed monthly fees predetermined based on the expected power consumption are applied for rural electrification projects and off grid bioenergy	• Feed-in tariff for wind, solar, ocean, run-river hydro and biomass • Fixed tariff for 20 years • Feed-in tariff allowance is also provided (uniform to consumers), 0.04057 PHP/kWh	No feed-in tariffs, but applicable payments and charges are applied for 1 MW$_{ac}$ embedded intermittent generation sources issued by energy market authority (EMA) 2015	• Earlier, adder scheme was applied and phase out in 2014–2015 • Feed-in tariff scheme has been started since 2015, for wind, solar, hydro, bioenergy (biomass, biogas and waste) • FiT premium[a] applied for bioenergy projects	• Avoided cost tariffs (ACTs) for small hydro, varied by year (time of use), by season and by region. The cost applied does not include the water resource tax, forest service fee and value added tax • Feed-in tariffs (FiT) are applied for wind (under revision), solar (COD before June 2019), MSW, and biomass (previously used ACTs) • FiT for biogas and geothermal are under study
Incentives	No incentives for RE, however, foreign investment law mentioned on	• Incentives under RA 9513 are income tax holiday and low	• Favourable tax incentives for all industry, including renewable energy	Thailand's Board of Investment (BOI)'s tax incentives for renewable	Corporate income tax: Rate 10% in 15 years, exemption for the first

(continued)

Table 2 (continued)

RE policy by category	Myanmar	Philippines	Singapore	Thailand	Vietnam
	income tax holiday, exemption, income tax relief, deduction for research and development and other duties exemptions	income tax rate, reduced government share, duty-free importation of equipment and VAT-zero rating, tax credit on domestic capital equipment, special realty tax rate on equipment and machinery, cash incentive for missionary electrification, exemption from universal charge, payment of transmission charges and tax exemption on carbon credits • Special incentives for mini-hydro and geo-thermal developers are also available: tax payable, income tax holiday and some exemptions.	• Singapore productivity and innovation credit offer 400% tax deduction (first SGD 400,000) or allowance and or payout for research in green innovation. 150% on expenditure in excess of SGD 400,000 • Market development fund for solar wind hydrogen and fuel cells, closed in March 2012. Energy research and development funds of SGD 25 million • Energy innovation programme and clean energy research Programme, each SGD 50 million. Energy training fund with SGD 20 million	energy: Income tax holidays up to 8 years, exemption or reduction of import duties on RE equipment, corporate income tax reduction	4 years and reduce to 50% in the next 9 years. Import tax exemption. Obligation to purchase electricity: All electricity generated from RE projects will be bought by EVN • Power projects under clean development mechanism (CDM) obtained more incentives on income tax, import duties, land rent and land use and subsidies • The compensation and supports for land clearances law on land
Financing supports	Off-grid RE based on rural electrification may	Government financial institutions shall provide	• SGD 20 million solar capability scheme • Solar developer may	• Energy conservation promotion fund: A source of venture	Under CDM project, some policies and

	have access to soft loans and grants	financial package for RE projects	obtain financial support up to 40% of the project cost (max. SGD 1 million per project)	capital for ESCOs to jointly invest with private operators in energy efficiency and renewable energy projects. The programme targets SMEs and smaller projects • Revolving fund: Provided via financial institutions for investment in energy efficiency improvement projects and renewable energy development and utilisation projects	financial mechanisms are applied
Permits and licences	No permits and licences	• One-stop shop for renewable energy (all processing application and contracts) • Energy virtual one shared system to handle and process service contracts in web-based platform • Evaluation process of renewable energy service contract (RESC). A competitive selection process (CSP) for all distribution utilities (DUs)	• Licencing framework for intermittent generation sources by EMA • Licencing for licenced electrical workers	• Energy industry act 2007 • Factory permit for small PV installation (< 1 MW) by ERC and Ministry of Industry	Power operation licences

(continued)

Table 2 (continued)

RE policy by category	Myanmar	Philippines	Singapore	Thailand	Vietnam
Technical standards	No technical standard applied	• Philippine grid code: required minimum connection, operational applicable for VRE, direction for VRE developers and manufacturers • Battery energy storage system (BESS) as a new source of frequency control ancillary services (FCAS) • Rules on interruptible load programme: addressing power shortage and augment limited power, minimising occurrence of manual dropping, incentivising and optimising resources, avoiding or minimising system emergencies, ensuring timely compensation and protecting public interest	Handbook for solar photovoltaic systems: singapore standard and codes for electrical installations as a safety for solar systems	• Adopted international standards, Thai industrial standards institutes applied national standards for solar energy • Grid codes from Electricity Generating Authority of Thailand – EGAT and provincial electricity authority – PEA • Codes of practices for RE projects (for PV and biomass) • Smart grid master plan	• Technical requirements for RE to National Power grid • Guidelines, rules, standards and procedures for maintenance, operation and development of Vietnam's transmission and distributions • Power grid operation circular consists of technical requirements for biomass, biogas and waste to energy, hydro wind and solar to power distribution

[a]Feed-in tariff (FiT) premium is the additional FiT set by the government which is only applied for the first 8 years of bioenergy project (such as biomass, biogas, MSW) in Thailand

4 Methodology on Collecting the Data

4.1 Country Experts Consultative Approach and Historical Review on ACE Databases and Studies

A comprehensive review from three published studies: Renewable Energy Policies in ASEAN, Renewable Energy Development 2006–2014 and Renewable Energy Outlooks for ASEAN was conducted as the main methodology for this chapter. Those studies were developed based on cooperation and consultation with the experts from ASEAN Member States. In addition, a further historical review from other ACE internal resources was also complementarily conducted to enrich the information for analysis. The resources were taken from ACE databases built from various ACE studies, workshops as well as country visits. This ACE visit's objective is to have in-depth discussion with experts and to ensure the data on policy/regulation and statistic are updated. The recent visits were conducted in January–February and June–July 2017.

4.2 REmap Analysis

Bottom-up approach from the energy balances was generated from REmap tools. The tools were developed by International Renewable Energy Agency (IRENA), which aims to calculate the amount of the selected renewable energy technology considering the economic parameters (in terms of costs both from government and developer perspectives), renewable energy potential and viability of the implementations from each ASEAN Member States. The matrices for the costs comprise of substitution costs, system costs, investment needs and renewable investment supports. In addition, externality analysis is also incorporated under this joint study between ACE and IRENA (ACE-IRENA 2016).

The substitutions costs are defined as the individual costs, while the specific technology (renewable or non-renewable technology) relatively is substituted with the specific non-renewable technology. It could be defined as the difference between the annualised cost of the REmap Options (USD/year) and the annualised cost of the substituted non-renewable technology (USD/year), divided by total renewable energy uses (GJ/year).

$$C_{\text{substitution}} = \frac{C_{\text{REmap}} - C_{\text{conventional}}}{E_{\text{REmap}}} \tag{1}$$

Substitution costs

Prior to set the substitution costs, the cost of technology for REmap Options (USD/year) shall be determined by summarising the annual capital expenditure

(USD/year), operating expenditures (USD/year) and fuel costs (USD/year). Fuel costs will be applied for the renewable energy with fuel supplies, like biomass.

$$C_{REmap} = C_{capital} + C_{operating} + C_{fuel} \qquad (2)$$

Costs of technology for REmap Options

In the second matrix, the capital investment cost (in USD/GW of installed capacity) in each year is multiplied with the deployment in that year or P_{RE} (GW) to arrive at total annual investment costs. The capital investment costs of each year are then summed over the period 2015–2025, so in conclusion $t_{REmap} = 15$ years (see Eq. 3). Net incremental investment needs in a country are the sum of the differences between the total investment costs for all technologies, renewable and non-renewable energy.

$$I_{needs} = \frac{P_{RE} \times C_{capital, GW}}{t_{REmap}} \qquad (3)$$

Investment needs

$$I_{country} = \sum_{i=1}^{n} I_{RE} - \sum_{i=1}^{m} I_{non\ RE}, i = \text{number of technology} \qquad (4)$$

Investment needs in one country

In addition, renewable investment support needs can also be approximated based on the REmap tool. Total requirements for renewable investment support in all sectors are estimated as the difference in the delivered energy service cost (e.g. in USD/kWh or USD/GJ based on a government perspective) for the renewable option against the dominant incumbent in 2030. This difference is multiplied by the deployment for that option in that year (E_{REmap}) to arrive at an investment support total for that technology (USD/year). The differences for all REmap Options are summed to provide an annual investment support requirement for renewables (ACE-IRENA 2016).

$$I_{support} = C_{subtitution, gov} \times E_{REmap}, \qquad \text{if } C_{substitution} > 0 \qquad (5)$$

Investment support for RE

4.3 Results on REmap

As the results, the specific increment of RE share for TPES, power, buildings and transport is suggested for each AMS towards the RE target of 23%. In terms of share, ASEAN shall reach 35% in power sector, 26% in buildings, 23% in Industry and 9% in transport sector. Figure 4 presented more detail contribution from each country in the sectors mentioned above.

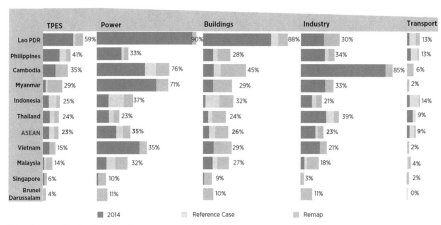

Fig. 4 Country RE share contribution in each sector to achieve RE target in ASEAN (ACE-IRENA 2016) (Notes: Reference Case is national target)

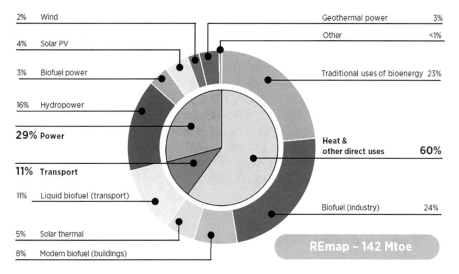

Fig. 5 Total renewable energy consumption in REmap (ACE-IRENA 2016)

In total, REmap suggests ASEAN to utilise 142 Mtoe of renewable energy in 2025 (Fig. 5). In more detail, 60% of renewable energy will be consumed in heat/ direct uses, 29% in power generation and 11% in transport sector. In power generation, hydropower will maintain its domination by contributing 16% of total energy consumption, solar PV at 4%, geothermal and biofuel power at 3%, wind at 2% and others for less than 1%. In industry, traditional uses of bioenergy and biofuel industries are utilised at 23% and 24%, respectively. In building, solar thermal is

ASEAN	Unit	2014	Reference Case 2025	REmap 2025
Total installed power generation capacity	GW	**195**	**387**	**422**
Renewable capacity	GW	51	124	180
Hydropower (excl. pumped hydro)	GW	39	79	82
Wind	GW	1	6	12
Biofuels (solid, liquid, gaseous)	GW	6	13	18
Solar PV	GW	2	13	55
CSP	GW	0	0	0
Geothermal	GW	3	10	11
Marine, other	GW	0	3	3
Non-renewable capacity	GW	144	263	242
Total electricity generation	TWh	**856**	**1656**	**1674**
Renewable generation	TWh	173	459	580
Hydropower	TWh	129	289	303
Wind	TWh	1	24	40
Biofuels (solid, liquid, gaseous)	TWh	22	54	81
Solar PV	TWh	2	19	81
CSP	TWh	0	0	0
Geothermal	TWh	20	59	67
Marine, other	TWh	0	9	9
Non-renewable generation	TWh	683	1202	1094

Left margin labels: Energy production and capacity / Power sector

Fig. 6 Composition of targeted AMS's RE installed capacity and power generation in REmap (ACE-IRENA 2016)

suggested to take a share around 5%, and the rest is biofuels (8%). In transport, the liquid biofuel supplied 11% of total energy consumption.

As shown in Fig. 6, in terms of power installed capacity, ASEAN should install around 180.2 GW of RE capacities of which hydropower and pump storage at 82.3 GW. It is followed by significant growth of solar PV, from only around 1.6 GW in 2014 to potentially increase to 54.7 GW. The significant growth will be affected by market growth increase, higher penetrations (more players), remarkably indigenous potential and its lower prices in the future. It is followed by bioenergy, winds and geothermal amounted to 41 GW. In total, ASEAN is expected to generate around 580 TWh electricity from renewables of which dominated by hydropower at 303 TWh, solar PV 80.7 and biofuel 80.8 GW.

5   Findings and Recommendations

Through this review, some findings and recommendations are highlighted as follows.

5.1   Dynamicity of RE Policy in ASEAN

In some AMS, the development of RE policies is rapidly changing. Out of the explanation whether this is good or not, the changes for the policy will take an implication to the market (developers, investor, consumer and the utilities company) directly. Recently, Vietnam and Indonesia changed their RE selling tariffs, whereas the Philippines and Myanmar are in progress of changing their RE targets.

In 2016, Indonesia promulgated a Feed-in tariff (FiT) for solar. Years before that, many FiTs were developed specifically based on the fuel types. In early 2017, all the FiTs were replaced by regional tariff based on production cost which determined under Ministerial Decree No. 12, and numbers of those tariffs were revised in the mid-2017 on Ministerial Decree No. 50. In the same time, Vietnam considered to revise the wind FiT and promulgated the FiT for solar in 2017. Furthermore, Myanmar is on the progress for reformulating their renewable energy target, and the Philippines (through Department of Energy or DOE) is on progress in amending new target for RE because of its solar target achieved in 2016. In the Philippines, solar PV remarkably grew from only 22 MW in 2014 to 142 MW in 2015 and 736 MW in 2016. The reason beyond this growth is the implementation of its attractive FiTs.

5.2 ASEAN's Country Level on RE Policy

In ASEAN RE policy study, AMS could be categorised into three categories or levels. These levels are (i) entry level, (ii) intermediate level and (iii) advanced level. In entry level, member states have no specific RE policy or still in the beginning of implementing some RE policies. In intermediate level, country awareness for the RE development and benefits was identified. Some policies were promulgated and applied but not at all. In terms of capacity, intermediate level country has solely installed huge amount of specific renewable energy type. In advanced level, countries have been applying all the six key policy factors and gained the significant renewable energy development from some of them. Some research has also been undertaken in those countries to promote their RE, and the installed capacity on them has been varied with many renewable energy sources.

Departing from those criteria, each AMS can be categorised to each level. Brunei Darussalam, Myanmar and Cambodia are categorised into the entry level countries. Lao PDR could be considered as the intermediate level country for its leading development in hydropower. Other five countries, namely, Indonesia, Malaysia, the Philippines, Thailand and Vietnam are considered as the advanced level country for the renewable energy policy in this region. Due to country geographical limitations, Singapore is also considered as an advanced level country since it has remarkably developed waste energy and solar, wide range of advanced research centres and activities, as well as established many attractive policies for renewable.

5.3 Feed-in Tariff for ASEAN

The Feed-in tariff (FiT) is one of the recommendations for entry level and intermediate level countries. The reasons are its high replicability and effectiveness in creating the market with attractive tariff for each electricity generated from renewable technologies. It is the one of the most implemented mechanism in the world (Ferroukhi 2012), and its success stories in ASEAN Member States has been proven like FiT story in the Philippines (for its solar PV with the results achieving their target), Malaysia (solar PV), Thailand (FiT and FiT Premium) and Indonesia (FiT is now based on production cost or BPP).

However, there are two disadvantages for FiTs that the country should be considered. First, the market created through this tariff is an artificial market. It means that the volume of the market is very much dependent on the changes (even drops) of the FiTs amounts. In the historical reviews, the member states who applied FiTs for more than 3 years tend to reduce the FiTs annually because as long as the volume of market expanded, it will be more burdening the national budgets. The way to solve this is by creating the soft transition from FiTs to the new tariff models (tariff based on production cost or market-based tariffs). This soft transition means to invite all the RE developers, consumers or other related stakeholders in discussing the shifting duration, how to determine the transition and what should they prepare and react to during each phase in transition. Their valuable feedbacks are very much necessary as the references of balancing the market characteristic and the funding limitations from the government.

As FiT provided by the government, the second disadvantage is its bond to the government's national budget. In FiT, the government made up the gap between real electricity price and renewable-based electricity tariffs. More projects will load the national budgets then, especially in the country where the electricity tariff and fossil fuels are subsidised. As the solution, FiTs also could be combined or in line with the subsidy reform programmes. The historical review found that some member states which have less/no subsidies on conventional energy may show quicker progress on the development for renewable energy, for example, Thailand and the Philippines. By having less or no subsidy, the gaps between FiT and real electricity tariffs (as baseline to set FIT) will be lower. As the results, the government burdens in supporting the FiTs will be reduced as well.

5.4 Advanced Incentives and RE Auctions

In addition, more attractive incentives (both in fiscal or non-fiscal) are the most recommended policies for all AMS, whether country is on the entry level or intermediate level or advanced level. Various types of incentives are still absent in the entry level AMS, and it has potential to be improved in future. For instances, in Myanmar there is no such incentives established to develop the renewable energy since the country has big potential of investment to replace the huge amount of biomass use with the renewable heating technology.

In fiscal incentives, the entry level countries could start to set more attractive incentives by improving their national investment law with the tax holiday, VAT deductions, exemptions or free charges for specific renewable energy projects or enterprises. Moreover, financing institutions in the entry level country are required to be trained and introduced with the benefits and risks profile as well as monetary basics of the renewable energy projects development to create friendly environment for early market development. These activities could be supported by knowledge and policy exchange, within ASEAN or from other advanced countries, who encountered many experiences in creating market for renewable energy, effectively. In addition, the expansions to other technologies are required for the intermediate country.

Competitive auctions by locations or projects are required to ensure the implementation of the project in the government planning. The competitive auction requires a comprehensive process flow and sufficient resources. Prior to implementing the auction, a geographical mapping for hydropower, solar PV, wind and biomass or other RE potentials are prerequisites for the country. Collaborations between the developers who own private databases or even its technologies and the government are strongly recommended in enriching knowledge databases for renewable energy potential.

In entry level country, the auctions shall be carefully conducted because government may have less information on the bidder experiences and status, competitiveness level of the market and appropriate pricing level for the winners. The preparatory actions to conduct an auction for entry level countries are (i) to learn the market status and potential bidders in their country as well as neighbouring countries; (ii) to conduct a comprehensive research on the available budget balances, projects volume and bidder price estimation in getting the results on market competitiveness and setting the winner pricing standards; and iii) to collect the comprehensive data and lesson learned from countries on the similar bidding contest. Although auctions may provide affordability and good quality projects, auctions will be ineffective if the tariff and incentives did not well set previously, like the case of Indonesia.

In Indonesia, geothermal is recorded only increased less than 500 MW in the last 7 years to be around 1.7 GW in the first half of 2017 (from 1.2 GW in 2010) even though the auctions and potential mapping are well conducted by the government (Directorate General for Electricity. Directorate General for Various New and Renewable for Energy Republic of Indonesia 2017). Geothermal has a high risk in its exploration due to its low probability and cost. The cost is ranged between US$ 3–8.5 million/well (Asian Development Bank 2015; Nathwani and Mines 2015) thus the electricity production cost from geothermal is still high. In developer's perspective, the exploration risks in geothermal are not compensable if the pricing of geothermal is following the rule of FiT and BPP (production cost) in the last 5 years. This fact shows that well-prepared auctions and mapping are not a guarantee to have the good results on renewable energy development. Analysing the suitable tariffs is more important to attract the developers. In addition, advanced technology development and the drilling and explorations guarantees from the government are necessary to compensate the high risk and high cost in geothermal exploration. Other activities such as looking at renewable energy portfolio or non-fiscal incentives, simplifying the land acquisitions and permit process, providing tax exemptions or other soft loans for geothermal development are also required to improve the effectiveness of auctions in RE development.

5.5   Grid Codes and RE Net Metering

Renewable energy integration required more technical standardisation and guidelines, as well as permit and licences. In entry level country which has less experience on setting the technical standards, the standards could be adopted into National Electricity Law. Under this law, the detail standards could be inherited to the specific topics like the grid codes or other technical standards. The consultative discussions with national utility and energy commissions/regulator bodies are requirements to ensure all related stakeholders involvement. In easier way, an adoption from international standards/guideline is another choice for entry level country in creating the national standards or guideline.

For intermediate and advanced level countries, each country has opportunities to improve and add new standards, guidelines or permit, and licences for specific renewable energy technology depend on needs.

As the grid codes and standards are established, both developers and utilities will find less issues on technical because the technical requirements are well met and harmonised with the utility standards. Furthermore, grid code and other standards will also encourage country to establish the net metering and net billing systems. This will encourage the consumers to be the prosumers (producer and consumer). If the net billing/metering is well created, then the dissemination to privates could be conducted after. The net metering and net billing could also be an important cornerstone for the country to establish the carbon taxes.

5.6   Carbon Taxes as a Regional Concern for ASEAN

Carbon taxes are another way to expand the renewable market where applied mostly after the other policies have already been applied. The implementation of carbon tax is very much recommended for the advanced and intermediate level countries; otherwise, it is not recommended for the entry level countries since the renewable energy market characteristics and players are not well identified yet. Carbon taxes are expected to skyrocket the renewable energy development (as well as energy efficiency) in revolutionary period.

However, this policy would have strong impacts to macrostructures markets and economics in country, for example, in economic growth, industrial developments, income distributions and international competitiveness. All of these effects can be minimised/ mitigated through an efficient recycling of tax revenues, an adoption of appropriate mitigating and an international agreement on a coordinated adoption of the tax (Cuervo and Gandhi 1998).

To implement a carbon tax programme, the penetration to selected sectors and related stakeholders is the key factor. The penetration is not only defined as internal penetration (national and state/provincial level) but also may be better to have it in external coordination (regional level or international level).

In internal penetration, the steps to implement the carbon taxes are at first, the dissemination of the importance points why the country shall apply the carbon taxes. At the same time, the government may collect the preliminary inputs from the stakeholders (experts, privates and industries) on how to provide some incentives (soft loans, tax exemptions), guidelines or other suitable policies for the tax objects, as mitigating actions to lower impacts of the taxes; in next steps, the government may formulate the policy and incentives; furthermore, the consultancy of those mitigating policies with some national stakeholders and experts on the related sectors are needed before its implementation; after the mitigating policies or regulations are fixed, the carbon taxes could be disseminated and applied in the country.

Prior to the implementation, an external coordination in international level or regional level on a coordinated adoption is suggested to be held in ASEAN. A consequence of the absence on its regional coordination is the competitiveness losses from the country who applies the carbon tax within the region. If carbon tax is applied only in one country, in the one competitive region, the country may take a risk of the opportunity losses in their business sectors, since the business developers will consider more than twice to develop their business in country with more taxes. This will result in the less competitiveness for the country, and the regional intentions are strongly recommended to overcome this.

There is a potential for AMS pursuing the collaborative actions to implement the carbon tax. These collaborative actions will be a cornerstone for AMS to achieve its RE target by 2025 as well as national intentions on the Paris Agreement. An agenda of setting regional efforts on carbon tax implementation will ensure this region to be avoided from the risks of carbon taxes.

5.7 RE Planning, Potential Mapping and Databases

In ASEAN, some AMS are still missing its database, electricity development planning and RE potential mapping. The establishment of electricity business plan like in Indonesia and Thailand is very important for the entry level countries. The improvement of data collections and capacity building for mapping the RE potential are required.

6 Conclusions

In REmap studies, each ASEAN Member State has their own contribution to achieve 23% RE target in their total primary energy mix by 2025. Power is one of the major contributors to achieve the RE target. However, the RE share in non-power is equally important to achieve the regional RE target. It is also recommended to ASEAN in considering the following action areas for enabling the ASEAN's RE potential: (1) by increasing power system flexibility while using renewable energy,

(2) by expanding efforts for renewable energy technology deployments, (3) by creating a sustainable bioenergy market and (4) by addressing the sharing of best practices for renewable energy. In addition, regarding to the varieties of the country's level in RE policy, ASEAN may provide a different treatment for each ASEAN Member States in developing their renewable energy both for power and non-power utilisations. For advanced level country, recommendations are the advanced incentives, subsidy reform and carbon taxes. In entry and moderate level, the establishment of Feed-in tariff, RE auctions and potential mapping is the ways to improve their RE market. If ASEAN goes with carbon tax, the member states shall do some regional collaboration on how to establish the carbon taxes in this region, by considering the macroeconomic effects. In conclusion, strengthening the collaborations on renewable energy sectors will effectively bring this region towards the 23% renewable energy target.

References

Asian Development Bank (2015) Unlocking Indonesia's Geothermal Potential. Manila. Accessed 23 Nov 2017. https://www.adb.org/sites/default/files/publication/157824/unlocking-indonesias-geothermal-potential.pdf

ASEAN Centre for Energy (2015a) 4th ASEAN Energy Outlook

ASEAN Centre for Energy (2015b) Asean Plan Of Action For Energy Cooperation (APAEC) 2016–2025. ASEAN Centre for Energy, Jakarta

ASEAN Centre for Energy (2016a) ASEAN Renewable Energy Policy. Jakarta

ASEAN Centre for Energy (2016b) ASEAN Renewable Energy Development 2006–2014. Jakarta

ASEAN Centre for Energy (2017) ASEAN Energy Databases on RE Policy and Regulations. Databases. Jakarta

Cuervo J, Gandhi V (1998) Carbon Taxes: Their Macroeconomic Effects and Prospects for Global Adoption - A Survey of the Literature. IMF Working Paper WP/98/73

Directorate General for Electricity. Directorate General for Various New and Renewable for Energy Republic of Indonesia (2017) The First Semester Achievement in 2017. Pers Conference. Jakarta, 4 Aug

Ferroukhi R (2012) Renewable energy tariff based mechanism. Presentation. IRENA

IRENA & ACE (2016) Renewable Energy Outlook for ASEAN: a REmap Analysis. International Renewable Energy Agency (IRENA), Abu Dhabi and ASEAN Centre for Energy (ACE), Jakarta. Asian Development Bank. 2015. *Unlocking Indonesia's Geothermal Potential*. Manila. Accessed 23 Nov 2017. https://www.adb.org/sites/default/files/publication/157824/unlocking-indonesias-geothermal-potential.pdf

Nathwani J, Mines G (2015) Cost Contributors to Geothermal Power Generation. Proceedings World Geothermal Congress 2015. Melbourne, 19–25 April

United Nations (2017) United Nations Sustainable Development Goals. Accessed 18 April 2017. http://www.un.org/sustainabledevelopment/energy/

Yosiyana B (2017) Overview of Renewable Energy Policy in ASEAN. Presentation. Bangkok, 11–13 July

Energy Sector in Malaysia: How Sustainable Are We?

Hoy-Yen Chan

1 Introduction

Reliable and sufficient energy supply will foster economic and social development. The modern energy services are crucial to produce food, provide and improve shelter, water, sanitation, medical care and education facilities and access to communication. All economy sectors, i.e. industry, commercial, residential, agriculture and transport, demand energy to raise productivity and generate local income. Energy industry in Malaysia plays a vital role of the national economic development, which contributed for about 20% of the total gross domestic product (GDP) in recent years (KeTTHA 2017). Though energy has become essential for the socio-economic activities, the energy sector is also the highest contributor to greenhouse gas (GHG) emissions. For instance, in 2011, it was 76% of the total emissions (NRE 2015). Therefore, it is important to ensure the energy sector is managed in a sustainable manner, which tackles the climate change problem. The objective of this chapter is to review and assess the sustainability of the energy sector in Malaysia. Energy statistics from using the national energy balance and reports are analysed. The energy production and consumption are evaluated from the perspectives of socio-economics and how this sector contributes to the GHG emission. This chapter starts from discussing the accessibility of modern energy services throughout the nation, which is one of the primary elements in ensuring a better quality of life. Then, the affordability is assessed by reviewing the energy subsidies. The discussion is followed by giving an overview of import and export trends of the fossil fuels since 1980 to recent years. In terms of the performance of the energy production and consumption, energy efficiency is evaluated for both supply and final consumption. Whereas, currently renewable energy is mainly fed into the electricity

H.-Y. Chan (✉)
Solar Energy Research Institute, University Kebangsaan Malaysia, Selangor, Malaysia
e-mail: hoyyen.chan@edu.ukm.my

© Springer International Publishing AG, part of Springer Nature 2018   25
H.-Y. Chan, K. Sopian (eds.), *Renewable Energy in Developing Countries*,
Green Energy and Technology, https://doi.org/10.1007/978-3-319-89809-4_2

supply and so is evaluated through the power generation mix. This chapter is then finalised with the overall GHG emissions that have been avoided and the potential in future years.

2 Access to Modern Energy Services

Energy is essential in human daily life that highly driven by the technologies. With limited access to energy, people may be living in poor conditions, which the implications are such as poverty and limitation of economic and human developments. For instance, electricity is a basic energy access in providing illumination, telecommunications and refrigeration that are the basic needs in our daily life. Therefore, it is important to measure the electrification rates of a nation. Malaysia has an overall electrification rate of about 98% in 2015, which the rates by regions were 99.9%, 95.1% and 94% for Peninsular Malaysia, Sabah and Sarawak, respectively (EPU 2015). During the implementation period of the 10th Malaysia Plan (2011–2015), rural electrification projects had benefited 115,153 housing units, mainly in Sabah and Sarawak through grid connection and renewable energy installations for the very remote areas. This in turns has increased the electrification rates from 84.4% and 72.1% in 2010 to 95.1% and 94% in 2015 for Sabah and Sarawak, respectively. As of 2015, only about a total of 2% of households in Malaysia did not have access to electricity, and it is targeted to achieve a 99% of national electrification coverage by 2020.

3 Energy Subsidies and Affordability

Energy prices could be an instrument for environmental and social costs to drive energy efficiency and renewable energy options. Furthermore, an appropriate pricing mechanism is important to ensure the energy affordability and subsidy fairness for different income groups. Thus, regulating appropriate energy prices is important in improving the social and economic development. The Malaysian government has been subsidising the commercial fossil fuels, and this could lead to energy intensive economy if it is not well structured. For example, though the regulated prices of natural gas have been revised, the current price of RM15.20/MMBtu for electricity subsector is still nearly RM33/MMBtu lower than the weighted average price in 2014. This has resulted in a cumulative subsidy amount of approximately RM227 billion. Subsequently, this also impacts on setting the electricity tariff, so as the LPG price structure, which is also heavily subsidised and has not been changed since 1983.

The burden of expenditure on fuel and electricity in household budgets is a measure of the affordability of commercial energy. This is important in justifying the local energy prices and tariff. The affordability indicators are compiled from the

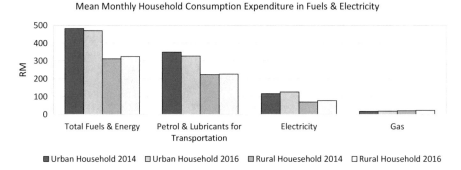

Fig. 1 Mean monthly household consumption expenditure in fuels and electricity in 2014 and 2016. (Data source: DOSM 2014b, 2016b)

Share of Monthly Household Consumption Expenditure on Fuels and Electricity (%)

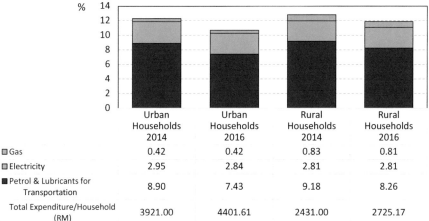

	Urban Households 2014	Urban Households 2016	Rural Households 2014	Rural Households 2016
Gas	0.42	0.42	0.83	0.81
Electricity	2.95	2.84	2.81	2.81
Petrol & Lubricants for Transportation	8.90	7.43	9.18	8.26
Total Expenditure/Household (RM)	3921.00	4401.61	2431.00	2725.17

Fig. 2 Share of monthly household expenditure on fuel and electricity in 2014 and 2016. (Data source: DOSM 2014b, 2016b)

surveys of household consumption expenditure and income by the Department of Statistics, Malaysia (DOSM 2014a, b, 2016a, b). Figure 1 shows the mean monthly household expenditure in fuels and electricity for the urban and rural areas in 2014 and 2016. These expenditures are about 11–13% of the total monthly expenditures (Fig. 2) and less than 1% different between the urban and the rural households. The highest spending of Malaysians in energy is the petrol for the transportation; however, the urban households have spent less in 2016 compared to 2014. This may due to the petrol prices that were changed from subsidised to managed float prices since December 2014, whereby the retail prices of RON95 and RON 97 were revealed on monthly basis and currently have been on weekly basis since April 2017. On the other hand, expenditure in electricity in 2016 has increased compared to 2014

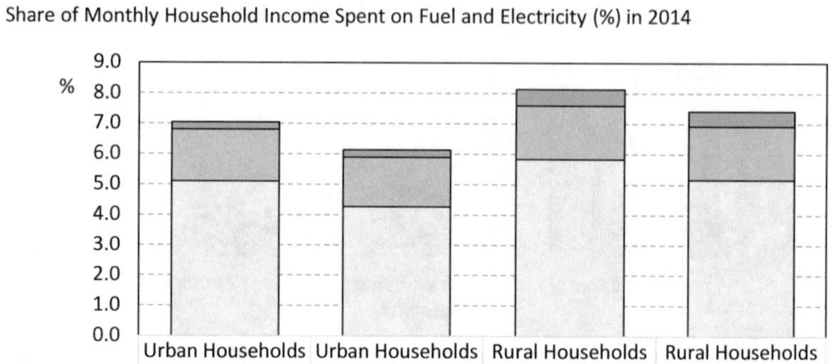

Share of Monthly Household Income Spent on Fuel and Electricity (%) in 2014

	Urban Households 2014	Urban Households 2016	Rural Households 2014	Rural Households 2016
▤ Gas	0.24	0.24	0.53	0.51
▣ Electricity	1.69	1.63	1.78	1.76
▢ Petrol & Lubricants for Transportation	5.11	4.26	5.82	5.16
Total Income/Household (RM)	6833.00	7671.00	3831.00	4359.00

Fig. 3 Share of monthly household income spent on fuel and electricity in 2014 and 2016. (Data source: DOSM 2014a, 2016a)

though there is no adjustment of tariff. This might be due to the heatwave that have caused a high temperature of near to 40 °C in March 2016, and thus higher electricity consumption for space cooling. As an overview, it was less than 1% different between the urban and the rural household expenditure in energy.

The average incomes per household per capita[1] of urban and rural in 2016 were about RM4,262 and RM2,564, respectively. The lower income groups usually spend a larger portion of their income on energy than do the rich. The households in rural spent about 7.4% of the total household income compared to 6.1% of the urban (Fig. 3). Nonetheless, in general the burden of fuels and electricity of the Malaysian society in 2016 was not the dominant, and commercial energy is considered affordable. This could be due to the subsidies on fuels and electricity that provided by the government. For instance, the monthly electricity bill that is RM20 or less will be waived; and the subsidy amount for residential electricity tariff is about RM0.08/ kWh, which the richer groups who consumed more electricity in fact received a bigger amount of this subsidy compared to the poor. On the other hand, for the rural households, renewable energy projects could be one of the contributions in reducing their burden on electricity.

It is anticipated that the next survey on shares of household consumption expenditures spent on fuel and electricity would be different from 2016. This is because the mass rail transit (MRT) in Klang Valley, which is the busiest region of the country, has started its services since July 2017. These might make changes in how the urban

---

[1]Estimated based on the average income per household and number of people receiving incomes from the (DOSM 2016a).

populations spend and commute in daily life. Therefore, this could reduce the expenses on petrol for transportation.

4   Present States: Reserves, Import and Export

Though Malaysia is one of the members of the Organisation of the Petroleum Exporting Countries (OPEC), their domestic productions of fossil fuels are declining. Moreover, oil and gas reserves though could last for more than 20 years, some of the fields are not feasible to be explored either they are small in size, scattered or remotely located. Therefore, additional supply needs to be secured to cater for future demand growth. Moreover, according to the report by the (IEA 2017), Malaysian gas production is going into a long-term decline after 2016, which would drop to about 20% below the 2016 production level by 2040. As a result, Malaysian LNG imports will continue to grow over the next 25 years to meet domestic demand. These can be seen from the recent trends of the fuels net export in Fig. 4. Only crude oil remains as a net export fuel, whereas natural gas has been imported since 2003 for the domestic supply, whereas coal is not an indigenous fuel. Natural gas and coal are the main resources for electricity generation in Malaysia; imports of these fuels that associated with the currency dependency and the current structure of the fuels subsidy are the challenges in keeping the electricity tariff low. In fact, this will impact on the overall economy of the country. The government needs to raise the awareness to the public that the resources used for electricity generation are imported, and the amounts have

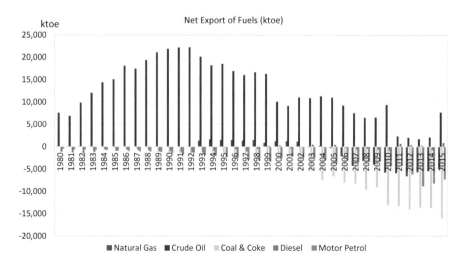

Fig. 4 Net export of fuels from 1980 to 2015. (Data source: ST 2017)

been increasing gradually. Similarly, motor petrol in transport sector is also a net importer, and diesel is not always a net exporter for the last 5 years.

5   Overall Use and Production Patterns

There are two common energy intensity indicators that reflect the energy use patterns of a nation, i.e. in terms of per capita and per unit of gross domestic product (GDP). The former is the energy use pattern of a society, whereas the latter is the trends of energy use related to the economic development. As shown in Fig. 5, in the case of Malaysian energy intensities, both the per capita and per GDP indicators have exhibited the similar trends for the energy uses in primary production, final consumption and electricity consumption. Thus, Fig. 6 is plotted to further understand these energy demands in relation to the population and economic growths. More efficient use of energy is whereby the energy uses are decoupled from the population and GDP growths. However, as shown in Fig. 6, both the primary energy supply and final energy consumption patterns are closely coupled with the GDP growth, whereas electricity consumption is relatively more sensitive to the number of population. Therefore, there are plenty of opportunities in energy efficiency improvement in Malaysia.

5.1   Energy Supply: Power Stations

The main types of power stations in Malaysia are hydro and thermal stations. The total conversion losses from the power stations in 2014 were about 22,785 ktoe, which was equivalent to about 64% of the total energy input of power generations. Thus, it is a need to look into the efficiency of the power stations to minimise the energy losses.

Hydropower
Though a hydropower plant is able to operate 24 h a day throughout the year, and with the current technology the highest efficiency of a hydropower plant is up to 0.95, hydropower plants in Malaysia only operate as peaking plants to meet the peak demand. Therefore, the ratio of electricity generated to energy input is not representing the power plant efficiency but rather the capacity factor, which is only about 0.38. Moreover, not only the remaining available capacity of hydropower has been reserved to comply a minimum reserve margin of 30% of the total national grid, the operation capacity also has been controlled to avoid the adverse impacts on the environment. Nonetheless, there is no official monitoring on the actual compared to the design operational performances of the hydropower plants. It was suggested that an important step such as institutionalising baseline performance indicators for

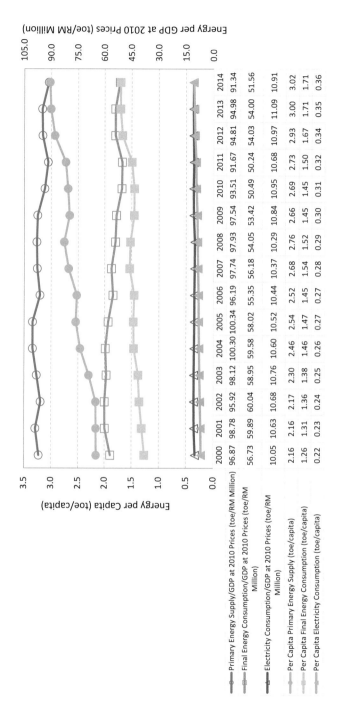

Fig. 5 Energy intensities indicators. (Data source: ST 2014)

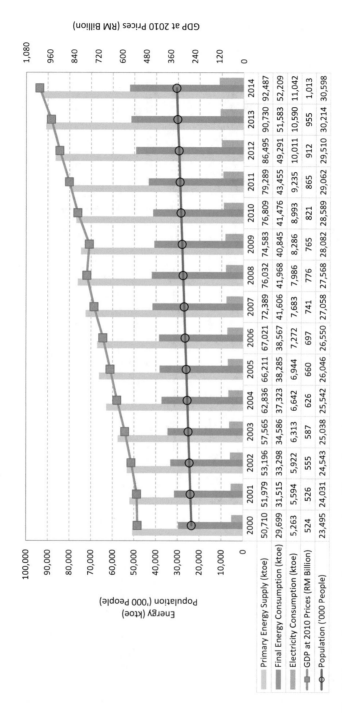

Fig. 6 Trends of energy supply, consumption, GDP and population. (Data source: ST 2014)

all hydropower plants in Malaysia based on the rated annual generation yields needs to be taken (ASM 2015). This hence would have a better performance monitoring system to maintain or improve the efficiency of the hydropower plants.

Thermal Stations

Continuous maintenance and heat losses reduction measures are crucial to ensure high efficiency of thermal power plants. It was acknowledged by the Energy Commission that the mode of operation, maintenance, ageing and degradation of the power stations are the key factors that will affect the thermal efficiency. For instance, by carrying out scheduled maintenance, the average thermal efficiency of coal thermal plants in Peninsular Malaysia was able to increase about 1% (ST 2015). This small percentage was in fact a significant saving in resources. A simple calculation performs here are based on an assumption of 1% improvement of the overall thermal efficiency in 2014. From the 2014 national energy balance, the average thermal efficiency of thermal stations was about 36%, and the losses were 19,514 ktoe. Therefore, without the 1% of efficiency improvement, the total losses would have to be 20,568 ktoe. This has given a total energy saving of 1,054 ktoe, which is equivalent to 12,258 GWh. This amount is nearly 8% of the total energy input from the coal and equivalent to annual electricity generated from a coal power plant that with the capacity of about 1500 MW. In this case, the 1% of efficiency improvement could give significant impacts, not only in terms of saving the consumption of coals that Malaysia is a net importer, could have saved the running costs of a coal power plant and reduced the carbon dioxide emissions.

5.2 Renewable Energy and Fuel Mix

Fuel mix in energy supply is an important indicator to determine the energy security and also the environmental impacts from the energy sector. As shown in Fig. 7, from 2000 to 2014, fossil fuels contribute a total of between 90% and 94% of total energy inputs in electricity generation. Natural gas is the dominant between 2000 and 2009; however, coal and coke have slowly become equally important energy sources since 2010 until the present. For the noncarbon energy sources, hydropower remains within 10% of the total generation due to the controlled capacity to converse the environment and also for the national grid reserve; and only about 0.5% is from other renewable energy sources such as solar, biomass and biogas. So, this gives a total of about 10.5% of renewable energy mix in power generation. However, these amounts are only for grid-connected generation; off-grid renewable energy electrification is not included. These indicate that the dependence on fossil fuels is very high though renewable energy has been introduced as the fifth fuel in the Five-Fuel Diversification Policy in 2000. Besides the natural gas that has been highly subsidised, other constraints in implementing renewable energy include high cost, stability of the supply, biomass feedstocks, lack of skilful personnel and financial aides. These issues have been acknowledged by the government, and solutions have been in planned especially in capacity building and also exploring new renewable energy

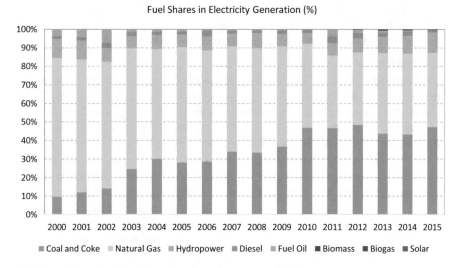

Fig. 7 Fuel shares in electricity generation. (Data source: ST 2017)

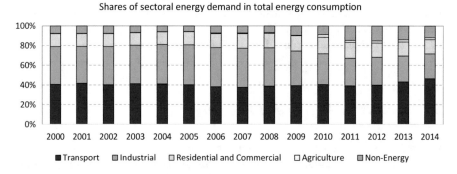

Fig. 8 Shares of sectoral energy demand in total energy consumption. (Data source: ST 2014)

resources such as wind, geothermal and ocean energy (EPU 2015). Actions need to be taken immediately in order to achieve the target of 20% from renewable energy in power generation mix by 2020 (KeTTHA 2017), which is less than 2 years from the present and a gap of about 8% to achieve.

5.3 Sectoral Energy Consumption

It is important to assess the sectoral energy consumption patterns and identify the potentials of energy efficiency measures and renewable energy applications. Figure 8 shows the shares of sectoral energy demand in total energy consumption. This is an

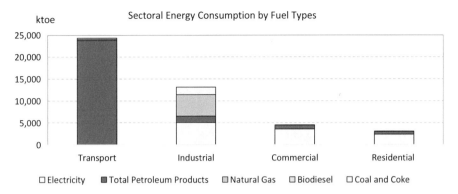

Fig. 9 Sectoral energy consumption by fuel types in 2014. (Data source: ST 2014)

energy indicator that shows the fractions of energy demand by different economic sectors. It provides information for policy makers to recognise which sector is the largest energy consumer. For the case of Malaysia, from 2000 until 2008, both transport and industrial sectors had almost equal energy demand; however, transport has been the dominant sector since 2009 until the present, and the energy demand ratio of transport to industrial sectors is approaching to 2 to 1. This indicates that in the period of 15 years, the growth rate of transport sector energy consumption has increased almost double the industrial sector. However, these two sectors use different types of fuels, and so different approaches and strategies will need to be formulated to shape the policy effectively. Therefore, major economic sectors are selected, and the energy consumption by fuel types in 2014 is plotted in Fig. 9. The petroleum products such as petrol and diesel are the main fuels consumed by the transport sector; natural gas and electricity are the main energy sources for industry productions; electricity remains the dominant for commercial and residential sectors.

Biodiesel in transport has a saving potential of 667.6 million litres of diesel a year (EPU 2015). Thus, in order to support the implementation of biodiesel in transport, 35 depots were constructed nationwide for blending facilities and would be able to encourage more biodiesel users and reduce the consumption of pure diesel in transport. Nonetheless, with two major interventions implemented, i.e. managed float to replace subsidised petrol prices and the mass rail transit (MRT) in Klang Valley, the energy consumption and the shares of fuel types in transport sector energy consumption in coming years may have changes if the public choose to take the public transport to commute. Furthermore, Malaysia has set an ambitious target of 100,000 EVs on the road by 2020 and 125,000 charging stations nationwide (IEA 2017).

For the non-residential buildings, initiatives that have been taken by the government are such as incorporating energy efficiency designs in new and retrofitting the existing buildings, setting air-conditioner temperature at a minimum of 24 °C, volunteering building ratings and revising the Uniform Building By-Law (UBBL) 1984 to incorporate the Malaysian Standard: Code of Practice on Energy Efficiency and Renewable Energy for Non-Residential Buildings (MS1525). The government has taken the lead whereby the government buildings become the models to demonstrate energy saving in buildings. For instance, four government buildings in Putrajaya were able to reduce the electricity consumption of 4% to 19% per month, which is equivalent to RM7,000 to RM130,000 savings (EPU 2015).

6 Overall Energy Efficiency

There is limitation in compiling the sectoral energy intensities over the corresponding value added. Therefore, sectoral energy efficiency analysis is not able to carry out with present available data. In regard to energy efficiency measures, apart from some initiatives that have been mentioned in previous sections, the government continues its effort in improving the energy efficiency in industry, non-residential buildings and household appliances. There are mainly on voluntary basis. Lack of enforcement could be one of the barriers in implementing energy efficiency measures. For instance, the government set the targets for a number of registered electrical energy managers and practice of ISO 50001 Energy Management System. However, no legal action or penalty on industries or commercial buildings that have exceeded the electricity consumption limit which was set under the provision of Efficient Management of Electrical Energy Regulations 2008, to ensure a continuous reduction in energy demand. On the other hand, by imposing penalty or providing incentives may enhance the motivations of moving towards energy efficiency. With the present situation, though the number of energy managers has increased, the demand for energy experts in energy management is relatively low. Market for the energy efficiency needs to be created through appropriate policy and regulations. Then, more energy efficiency equipment and technologies would be brought into the market and so as the business and job opportunities. Furthermore, to date, the energy efficiency strategy in residential sector is only focusing on four major appliances, i.e. air-conditioner, refrigerator, fan and light. There is no minimum standard or reference with respect to energy performance for the dwellings in Malaysia. With the increasing number of urban population, the demand for residential buildings will be increased. Thus, it is important to study the energy efficiency designs potential in reducing the dwellings energy consumption.

7 GHG Emissions

The total GHG emissions in 2011 were 290.23 Mt. CO_2eq, which 72% of the emissions are CO_2, and the major emitter was from energy industry followed by the transport sectors. The emissions have been reduced gradually since 2005 after some mitigation interventions have been implemented. As of 2011, the total emission avoidance from the energy sector was about 2487 Gt CO_2eq. Through the

Table 1 Emission reduction in 2013 and potential reduction by 2020

Mitigation action from the energy sector	Emission reduction achieved in 2013 (kt CO_2eq)	Potential emission reduction in 2020 (kt CO_2eq)
RE implementation through feed-in tariff mechanism	252.78	5458.09
RE electricity generation by non-feed-in tariff-regulated public and private licensees and other mechanisms	948.77	2179.29
Use of palm-based biodiesel in blended petroleum diesel	719.74	1802.49
Application of green technology	94.81	1426.35
Implementation of green building rating scheme	60.40	858.40
Efficient electricity consumption in all federal government ministry buildings (baseline established in 2013)	–	98.21
Reducing emissions through development and usage of energy-efficient vehicles (EEVs)	40.96	199.74
Use of compressed natural gas (CNG) in motor vehicles	154.62	217.57
Rail-based public transport	214.93	977.51
Total	2487.01	1,3217.65

Source: NRE (2015)

SAVE programme that was implemented from 2011 to 2013, a total of 306.9 GWh was saved from more efficient refrigerators, air-conditioners and chillers. This amount of energy saving is equivalent to 208,705 tCO_2eq GHGs avoidance. On the other hand, in 2014, biodiesel has been introduced nationwide to reduce the consumption of commercial diesel in the transport sector. The blended B7 biodiesel has managed to reduce GHGs emission by 1.7 million tCO_2eq. In addition, other strategies such as retrofitting government buildings, recycling of solid wastes and development of green products and forest reserves have contributed a total avoidance amount of at least 16 million tCO_2eq of GHGs. Nonetheless, the achievements have come along with policy interventions through renewable energy feed-in tariff, green building indexing, green technology financial scheme and relevant roadmaps and guidelines. The details achievements and projections of CO_2 reductions are shown in Table 1 (NRE 2015).

Under the United Nations Framework Convention on Climate Change (UNFCCC), the Paris Agreement was adopted on 12 December 2015 to gather the global efforts on combating the climate change. The ultimate aim of the agreement is to keep a global temperature rise below 2 °C above pre-industrial levels and whenever possible limit the temperature increase to 1.5 °C. A total of 160 countries have ratified the agreement, and it was entered into force on 4 November 2016.

Malaysia is one of the countries who commit to reduce the greenhouse gas (GHG) emissions, with an Intended Nationally Determined Contribution (INDC) of:

> Reduce its greenhouse gas (GHG) emissions intensity of GDP by 45% by 2030 relative to the emissions intensity of GDP in 2005. This consist of 35% on an unconditional basis and a further 10% is condition upon receipt of climate finance, technology transfer and capacity building from developed countries. (The Government of Malaysia 2015)

Therefore, the next step for the Malaysian government to ensure further GHGs emission reduction is to strengthen the financial mechanism in management of natural resources, including the design of public payment methods, for instance, carbon offsets and carbon tax could be the options to be considered.

8 Conclusions

The review and assessment exercises can be concluded with summaries and recommendations as follows:

Highlights and Summary

- Malaysia has an overall electrification rate of about 98% in 2015.
- The total spending of fuels and electricity of a household in 2014 is near to 13% of the total household consumption expenditure.
- The subsidy amount for residential electricity tariff is about RM0.08/kWh, which the richer groups who consumed more electricity in fact received a bigger amount of this subsidy compare to the poor.
- Cumulative subsidy for natural gas until 2014 is approximately RM227 billion.
- Natural gas has been imported since 2003, and the import will continue to grow over the next 25 years to meet domestic demand.
- Both the primary energy supply and final energy consumption patterns are closely coupled with the GDP growth.
- The growth of electricity consumption is relatively more sensitive to the number of population.
- By increasing 1% of the thermal efficiency of coal thermal plants, this has saved approximately 1,054 ktoe of energy or equivalent to about 8% of the total energy input from the coal.
- As of 2015, it is about 11% of power generation mix that was from the renewable energy sources.
- Transport sector has been the dominant energy consumer since 2009.
- The total GHG emissions in 2011 were 290.23 Mt. CO_2eq, which 72% of the emissions are CO_2, and the major emitter was from energy industry.
- As of 2011, the total emission avoidance from the energy sector was about 2487 Gt CO_2eq.

Recommendations

- The energy subsidies need to be structured to continue to benefit the poor but not the rich income groups.
- The government needs to raise the awareness to the public that the main resources used for electricity generation are coal and natural gas, which they are imported fuels.
- The government needs to raise the awareness to the public that motor petrol in transport sector is also a net importer, and diesel is not always a net exporter for the last 5 years.
- The energy intensity indicators that coupled with GDP and population showed that there are great opportunities for energy efficiency.
- Institutionalising baseline performance indicators for all hydropower plants in Malaysia that based on the rated annual generation yields.
- Continuous maintenance and heat losses reduction measures to ensure high efficiency of thermal power plants.
- Actions need to be taken immediately in order to achieve the target of 20% from renewable energy in power generation mix by 2020, yet a gap of about 8% to be achieved in 2 years' time.
- Lack of enforcement could be one of the barriers in implementing energy efficiency measures.
- By imposing penalty or providing incentives may enhance the motivations of moving towards energy efficiency.
- Shaping policy that would create market for green technologies.
- Establish a reference or guideline with respect to minimum required energy performance for the dwellings in Malaysia.
- Strengthen the financial mechanism in management of natural resources, including the design of public payment methods, for instance, carbon offsets and carbon tax could be the options to be considered.

References

ASM (2015) Carbon free energy: roadmap for Malaysia. Academy of Sciences Malaysia, Kuala Lumpur

DOSM (2014a) Household income and basic amenities survey report 2014. Department of Statistics Malaysia, Putrajaya

DOSM (2014b) Report on household expenditure survey 2014. Department of Statistics Malaysia. Available at: https://newss.statistics.gov.my/newss-portalx/ep/epProductForm.seam?cid=19185#

DOSM (2016a) Household income and basic amenities survey report 2016. Department of Statistics Malaysia. Available at: http://www.statistics.gov.my/portal/index.php?option=com_content&view=article&id=1640&Itemid=111&lang=bm

DOSM (2016b) Report on household expenditure survey 2016. Department of Statistics Malaysia, Putrajaya

EPU (2015) Stratergy paper 17: sustainable usage of energy to support growth. In: Eleventh Malaysia plan 2016–2020: anchoring growth on people. Economic Planning Unit Prime Minister's Department Malaysia, Putrajaya, p 38

IEA (2017) Southeast Asia energy outlook 2017. International Energy Agency, Paris

KeTTHA (2017) Green technology master plan Malaysia 2017–2030. Ministry of Energy Green Technology and Water, Putrajaya

NRE (2015) Malaysia biennial report to the UNFCCC. Ministry of Natural Resources and Environment, Putrajaya

ST (2014) National energy balance 2014. Suruhanjaya Tenaga (Malaysian Energy Comission), Putrajaya

ST (2015) Performance and statistical information on electricity supply industry in Malaysia. Suruhanjaya Tenaga (Malaysian Energy Commission), Putrajaya

ST (2017) National energy balance 2015. Suruhanjaya Tenaga (Malaysian Energy Comission), Putrajaya

The Government of Malaysia (2015) Intended nationally determined contribution of the Government of Malaysia. Available at: http://newsroom.unfccc.int/unfccc-newsroom/malaysia-sub mits-its-climate-action-plan-ahead-of-2015-paris-agreement/

Renewable Energy in Achieving Sustainable Development Goals (SDGs) and Nationally Determined Contribution (NDC) of Vietnam

Ha Ninh Tran

1 Overview of Renewable Energy in Vietnam

Energy always is a very important element for the development of each country. The more developed the society, the higher the demand for energy. However, the traditional source for it, fossil energy, is gradually depleted in proportion to the pace of economic development in the world. Therefore, conflicts, civil and regional wars, and hostilities in recent years have been related to energy directly or indirectly. In the past time, traditional energy sources have been exploited mainly for oil, coal, gas, and electricity, but in the next few years, these sources of energy will be depleted soon. In other words, they cannot be sustainable sources for future development. Figure 1 shows the shares in power generation by fuel types as of 2016.

Meanwhile, the waste of energy in industrial production, construction, and transportation and households in the world as well as in Vietnam is quite common for that the global energy demand is increasing rapidly. It is predicted that Vietnam will only meet 30% of its domestic energy consumption by 2025 and Vietnam will likely have to import energy in the future. Against that prediction, the shift from fossil fuels to cleaner energy such as renewable energy (RE) is becoming increasingly urgent (Huong 2012).

Vietnam is considered to be one of the countries having high potential for renewable energy, which are occurred all over the country. Figure 2 shows the current use and total potential of each renewable source in Vietnam. For examples, the estimate biomass potential from agricultural residues or wastes that can produce about ten million toe per year. Approximately ten billion cubic meters of biogas can be obtained from domestic wastes, husbandry, and agricultural residues. The abun-

H. N. Tran (✉)
Department of Climate Change, Ministry of Natural Resources and Environment of Vietnam, Hanoi, Vietnam

© Springer International Publishing AG, part of Springer Nature 2018
H.-Y. Chan, K. Sopian (eds.), *Renewable Energy in Developing Countries*,
Green Energy and Technology, https://doi.org/10.1007/978-3-319-89809-4_3

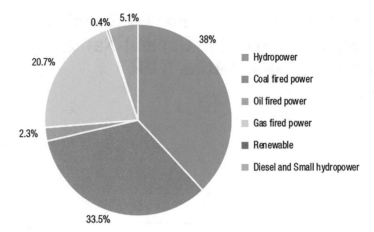

Fig. 1 Power generation by fuel types in Vietnam as of December 31, 2015. (Source: EVN 2017)

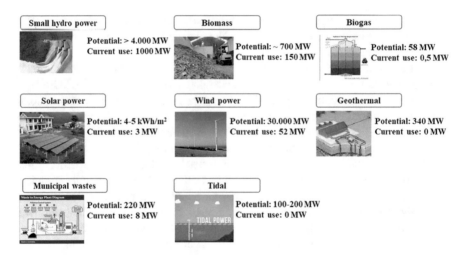

Fig. 2 Potential and current use of renewable energy in Vietnam. (Source: Huong 2014; Cuong 2012; Vietnam's National Agency for Science and Technology Information 2015)

dant solar energy with average solar radiation is 5 kWh/m²/day. In addition, having geographically more than 3400 km of coastline, Vietnam has a great potential for wind energy, estimated at 500–1000 kWh/m²/year. These RE sources will be used to meet the rapid-increasing energy demands. The overview of RE sources in Vietnam is as follows (Huong 2014; Cuong 2012; Vietnam's National Agency for Science and Technology Information 2015):

Small Hydropower
Looking at the contribution structure in the electricity sector, hydropower is still taking a very large share. However, electricity output from hydropower plants is often unstable due to the dependence on the water flow and the amount of water

stored in the reservoirs. With small hydropower, up to now, about a half of the potential has been exploited; some of the remains have been located in remote and/or unfavorable areas which have high exploitation costs. According to the recent assessment reports, over 1000 sites have been identified that have the potential for small hydropower development, ranging from 100 kW to 30 MW with a total installed capacity of over 7000 MW, mainly located in the Northern Mountainous, South Central and the Central Highlands areas in Vietnam (Huong 2014; Cuong 2012; Vietnam's National Agency for Science and Technology Information 2015).

Wind Power

Located in the subtropical monsoonal region with long coastline, Vietnam has a fundamental advantage for exploiting wind energy. Comparing the average wind speed in the East Sea areas of Vietnam with other close-by areas, it is clear that the winds in the East Sea are quite strong and change with the season. Under the Energy Assessment Program in Asia, the World Bank has been carried out a detailed survey of wind energy in Southeast Asia, of which Vietnam has the largest wind potential with total potential of wind power estimated at 513,360 MW. That is more than 200 times the capacity of Son La hydropower and more than 10 times the total forecasting capacity of the power sector by 2020. Obviously, it is still a long way to go from the theoretical potential to the exploitable potential, to the technical potential, and finally into the economic potential (Huong 2014; Cuong 2012; Vietnam's National Agency for Science and Technology Information 2015).

Biomass

With the advantage of an agricultural country, Vietnam has a huge and varied biomass of wood, rice husk, coffee grounds, straw, and bagasse. Agricultural residues are plentiful in the Mekong Delta, accounting for about 50% of total national agricultural residues and the Red River Delta with 15% of the country's total residues. There are annually nearly 60 million tons of biomass from agricultural residues, of which 40% is used to meet household energy needs and electricity production. Other biomass resources include wood products, urban waste, and animal husbandry waste. Wood products and chips from wood processing companies are originating from natural or planted forests and imported timber. At present, 90% of biomass production is used for cooking. Biomass is used in two areas including heat and electricity production. For heating, biomass supplies more than 50% of the total primary energy consumed in Vietnam. Biomass has been declining in recent years as other modern forms of energy such as liquefied petroleum gas (LPG), etc., have been put to use. In rural areas, biomass is still the main source of fuel for cooking in more than 70% of the rural household. Biomass is also a traditional fuel source for many local manufacturing plants such as food processing, fine art, and brick and ceramic production (Huong 2014; Cuong 2012).

In addition to meeting the energy needs, using appropriate biomass will also help to reduce greenhouse gas (GHG) emissions, minimize health damage caused by burning wood and charcoal, reduce poverty, and improve sanitation.

Solar Energy

Vietnam is considered as a country with high potential on solar power, especially in the central and southern regions of the country, with an average solar radiation intensity of about 5 kWh/m^2. Meanwhile, the intensity of solar radiation is lower in the northern regions, estimated at 4 kWh/m^2 due to weather conditions due to the cold and cloudy weather in winter and spring. In Vietnam, average solar radiation of 230–250 kcal/cm^2 with the direction to increase southward is about 2000–5000 light-time hours per year, with estimated theoretical potential of about 43.9 billion toe. Solar energy in Vietnam is available year-round, quite stable and widely distributed across different regions of the country. In particular, the average number of sunny days in the central and southern provinces is about 300 days/year. Solar energy is used mainly for purposes such as electricity production and heat supply (Huong 2014; Cuong 2012).

Geothermal

Geothermal energy has been used to burn and bathe since ancient time, but today it is used to generate electricity. Approximately 10 GW of geothermal power capacity was installed in the world by 2007, providing 0.3% of global electricity demand. In addition, 28 GW of direct geothermal capacity is installed for heating, spa, industrial processes, seawater filtration, and agriculture in some areas.

Exploitation of geothermal energy is economically feasible and environment friendly but was previously geographically limited to areas near plate tectonic boundaries. Recent scientific and technological advances have gradually expanded the scope and area of these potential resources, particularly direct applications for household heating. Geothermal wells tend to emit GHG trapped underground, but these emissions are much lower than fossil fuel combustion. This technology is likely to help mitigate global warming if it is widely deployed (Huong 2014; Cuong 2012).

Although geothermal resources have not been investigated and calculated. However, the latest survey and assessment data show that the geothermal potential in Vietnam can be exploited to over 300 MW. It can be effectively exploited in the Central area of Vietnam.

Tidal

Although Vietnam has a very long coastline, wave and tidal energy sources have not made a significant contribution to the system due to the low investment. Up to now, many wave and wave designs are available; however, they are all in the testing phase. These designs have the advantages of not affecting landscapes and coastal and offshore ecosystems like offshore wind farms.

The advantage of tidal power projects (in comparison with wind power and hydropower) is that it is possible to forecast tides, thus predicting their capacity. Until now, traditional tidal use in Vietnam has been limited to use for salt production and regulate irrigation for aquaculture.

In general, tidal power potential of Vietnam is not large compared to other countries in the world. Vietnam's tidal power reserve is estimated at 1.6 billion kWh per year and concentrated in the coastal area of Quang Ninh province (about

1.3 billion kWh per year). Approximately 0.2 billion kWh/year can be exploited with small capacity in the downstream of Mekong river (Vietnam's National Agency for Science and Technology Information 2015).

At present, some of RE sources in Vietnam have been exploited to produce electricity. According to incomplete statistics, their total installed capacity is about 1215 MW, in which small hydropower (1000 MW), biomass (152 MW), municipal waste (8 MW), solar (3 MW), and wind (52 MW). The current exploitation of RE is very small compared to their potential. Given that context and forecasts in the coming time, there should be specific solutions to increase the level of RE development.

2 Renewable Energy in Achieving the Sustainable Development Goals (SDGs)

Sustainable development has become the strategy and vision of the Parliament and the Government of Vietnam and has been integrated and reflected in the various strategies, plans, programs, and projects of Ministries, economic sectors, and localities for socioeconomic development and environment protection. Many related policies were promulgated to ensure sustainable development goals and the implementation of international agreements to which Vietnam is a party (Ministry of Natural Resources and Environment of Vietnam 2014). In order to implement SDGs of the United Nations 2030 Agenda adopted by the General Assembly of the United Nations in September 2015 and to promote the national socioeconomic development, the Prime Minister of Vietnam's Government issued a National Action Plan for implementing the 2030 Agenda for Sustainable Development identified 17 SDGs of Vietnam by 2030 (The Prime Minister of Government 2017a). The overall objective of this Action Plan is to "Maintain sustainable economic growth coupled with the promotion of social and equitable advancement and ecological environment protection, management and effective use of natural resources, and proactive response to climate change; ensure every individual fully develops their potential, participates in and equally benefits from development achievements; and thus build a peaceful, prosperous, inclusive, democratic, equitable, civilized, and sustainable Vietnam society" (Ministry of Natural Resources and Environment of Vietnam 2017a). This Action Plan is one of the great efforts in the formation and development of the institutional system in accordance with the requirements for sustainable development. It is expected that RE development will help to achieve the targets which have been set under this Action Plan with their great potential.

2.1 Direct Contribution

In Vietnam, the demand for electricity has been increasing rapidly to meet the demand of economic and social development. The electricity production output despite a huge increase but still significant shortfall amounts, especially in the dry season, when the amount of water needed for hydropower, is insufficient due to lack of rain. This has led to the phenomenon of power failure in some local areas which cause negative impacts on the living conditions of the people and production. In order to offset the shortage of electricity, in recent times, Vietnam has started to import electricity, which is not to mention the import of coal every year to cater to the thermal power plant. The concerns about energy security are growing in the country.

Fossil fuel production is declining due to limited reserves and difficult conditions to exploit, but the energy demand is growing and growing. In addition, the consumption of fossil fuel is daily emitting large amounts of GHGs and causing serious environmental pollution. Meanwhile, the potential of new and RE sources is considered to be huge; the development of RE will contribute to reducing fossil fuel consumption which will reduce GHG emissions. Electricity generation from RE sources is therefore considered to be the ideal supplement to power shortages and not only helps to diversify energy sources but also contributes to divide risks, enhance and ensure national energy security, and provide higher living standards, especially in remote and island areas, especially where the transmission of national grid cannot be reached.

According to statistics of the Electricity of Vietnam (EVN), by 2016, the national electricity grid has supplied electricity to nearly 99% of households in Vietnam (EVN 2017). The remaining, about one million people, are located in remote areas, where infrastructure and economic conditions are still low. Lack of access to electricity grid has become a barrier for economic development; moreover, this may also lead to deforestation for crop cultivations. If people, especially ethnic minorities are provided a stable, affordable power source, it will enable people stabilize their life and development of production; access to knowledge, which deliver the opportunities to change their structure and type of production; increase the local economic development, stable life, but do not depend on the forest; and reduce the shifting cultivation.

In recent years, in some remote and isolated areas, there have been a number of projects using RE such as wind and solar power projects to take advantage of the available natural conditions of those areas and provide the electricity for people. The RE technologies are expected to provide more alternative energy sources to users, especially those in remote areas where national electricity grid is inaccessible. In addition, with the current technology, the integration of RE solutions in urban construction such as the installation of solar panels on roofs or small-scale wind power plants is also possible, which is expected to reduce the pressure on the grid during the peak time in every year.

In the past, concerns about investment costs and supporting services conditions were a great barrier, making RE technologies difficult to apply, but now with the development of technology, the cost of electricity production from RE has declined sharply. For example, the cost of producing solar electricity has dropped to just one-third in the period of 2009–2016, and wind power production price has also fallen to around US$ 80/MWh. Wind and solar power are the two most potential sources of RE in Vietnam. It can be seen that Vietnam is increasingly getting closer to a future where RE production such as wind and solar power is less expensive than from fossil fuels (Frankfurt School-UNEP Centre/Bloomberg New Energy Finance 2017).

Recognizing the important role of RE in the development of the country, the Government of Vietnam has issued a series of policies to encourage RE development, of which the most important is the RE development strategy in Vietnam to 2030, with an outlook up to 2050. This strategy provides an overall vision with a long-term vision for the development of RE sources in Vietnam (The Prime Minister of Government 2015). The development strategy aims to:

- Encourage and mobilize all resources from the society and people for RE development
- Ensure better access to modern, sustainable, reliable, and affordable energy sources by all citizens
- Accelerate the expansion and use of RE sources
- Increase the domestic energy supply
- Gradually increase the RE share in the national energy production and consumption in order to ensure less dependence on fossil sources
- Contribute to better energy security, mitigating climate change, environmental protection, and sustainable socioeconomic development (Ngan and Huong 2016)

The Government of Vietnam has also issued several policies to support some individual RE sources such as wind power, solar power, etc., to promote and attract investment to develop RE sources which are easy to exploit and high potential. For example, in April 11, 2017, the Prime Minister of Vietnam issued the encouraging mechanisms for solar electricity generation development. In which, in the period from June 2017 to the end of June 2019, individuals and organizations have involved in the development of solar electricity generation projects that are legally mobilized from domestic and/or international individuals and organizations in accordance with the current regulations. The project developers are exempted from import tax for imported goods to create fixed assets for the project, which exempted and reduced the corporate income tax. Commercial solar electricity generation projects will be purchased at VND 2086/kWh (equivalent to US cent 9.35/kWh) (The Prime Minister of Government 2017b). Although these incentives are mostly relatively short lived, only last for about 2 years, from 2017 to 2019, these policies have generated the great momentum to create an energizing market for solar power development in Vietnam.

Despite many challenges still occur, with potential as well as the orientation on RE development in the coming time, RE will become one of the key motivations for socio-economy development in Vietnam and contribute to ensure universal access to modern energy services, improve efficiency, and increase the use of RE sources (SDG No. 7). RE will help to reduce fossil fuel use and thereby contribute significantly to reduce environmental pollution and GHG emission. The steady growth in deployment of renewables, the spread of smart energy technologies such as efficient lighting, and the softer-than-expected trend on electricity demand are limiting the growth of world energy sector emissions. The UN Environment's Emission Gap Report 2016, published last November, said: "In 2015 global CO_2 emissions stagnated for the first time and showed signs of a weak decline compared to 2014 (of 0.1%). This was preceded by a slowdown in the growth rate of CO_2 emissions, from 2% in 2013 to 1.1% in 2014." However, the same report also warned: "The world is still heading for a temperature rise of 2.9 to 3.4°C this century, even with Paris pledges." Some individual countries have performed well recently in terms of emission reduction. For example, the UK total net CO_2 emissions were 383.8 megatons in the year to the second quarter of 2016, down 29% from 2007 and 36% from the peak year of 1991 (Department for Business, Energy, and Industrial Strategy 2016). In the case of China, the International Energy Agency said in March 2016 that emissions dropped 1.5% in 2015, defying the agency's prediction from 2010 that Chinese emissions would grow 1.6% per year between 2008 and 2035 (International Energy Agency 2016). However, forecasts on global emissions are bleak. Most expect rising electricity demand in emerging economic regions such as India and South East Asia to lead to greater coal-fired generation and to higher CO_2 output (Frankfurt School-UNEP Centre/Bloomberg New Energy Finance 2017).

As one of the most vulnerable countries to climate change, Vietnam has been actively promoting internal resources, calling for international support to carry out activities to respond to climate change with the orientation to develop a low-carbon economy, green growth, and sustainable development and actively cooperate with the international community to achieve the ultimate goal of the UNFCCC and keep the global average temperature at no more than 2 °C by the end of this century (Ministry of Natural Resources and Environment of Vietnam 2014).

In 2015, to contribute to the process of the Paris Accord, Vietnam announced its intended nationally determined contribution (INDC) with the voluntary emission reduction of 8% compared to the business-as-usual (BAU) scenario and will increase to 25% if having fully supported internationally. In order to meet these emission reduction targets, a series of specific emission reduction policies and plans are proposed, in which RE will play a crucial role. Thus, the development and application of RE technologies will contribute to Vietnam's GHG mitigation, contribute to urgently act to combat climate change and its impacts (SDG No. 13).

2.2   Indirect Contribution

In addition to the direct contributions mentioned above, RE has the potential to make a significant contribution to the achievement of other sustainable development goals indirectly. Overlooking the issues related to economic viability, RE development will provide diversified alternative energy sources to people, especially in remote areas, where the national electricity grid is difficult to be reached. The electricity from RE sources will be provided for economic development and diversification of industries in these areas. Moreover, RE will help create more employment opportunities and create new livelihoods, which will increase their incomes. Electricity will help people to enhance production processes and increase productivity. The food production and food processing are boosted based on the availability and stability of electricity (due to the availability of alternative sources of renewable energy, independent of national grid). The productivity increased will contribute to hunger eradication and poverty reduction for local people, contributing to meeting basic nutritional needs of people (SDG No. 1 & 2).

The diversification and stability electricity generated from RE source will also contribute to develop and improve the health-care infrastructure system for people in local remote areas, to reduce the over-loading situation on the central hospitals in the cities. Thus, people will have it easier to access health-care services at local (SDG No. 3). In addition, electricity provided will help people improve the conditions of education and living standards in remote areas. Their education system has been developed to raise the level of knowledge and education (SDG No. 4).

With the application of smart grid technology, RE will be maximized in use and ensure the continuous and stable supply of electricity for economic activities. The application of RE under a smart grid will meet the energy needs of the people and almost certainly not create the bad consequences for the environment. Consequently, RE will help Vietnam to meet economic development goals without compromising or minimizing environmental pollution compared to fossil fuels and to develop a sustainable production economy (SDG No. 8). Using RE in production and consumption as alternatives for fossil fuels will reduce a large amount of GHG emissions because RE technologies are mostly zero emissions or low emissions. For that reason, it will help to reduce the amount of GHG emissions by gross domestic product (GDP) with relative to the contribution of RE in the economy. This is expected to help promote a comprehensive and sustainable industrialization process toward the development of a green and environmentally friendly economy in Vietnam (SDG No. 9).

As an agricultural country with almost 70% of the labor force concentrated in rural areas, agricultural production always accounts for a large proportion of food production reaching nearly 50 million tons (General Statistics Office 2016). However, this also has been given a side effect, annual huge amount of agricultural residues. Without treatment solutions, they will become a source of environmental pollution, extremely high GHG emissions. Particularly, when the farmers in areas such as suburban districts of Hanoi and some localities burn them in fields, they will cause the air pollution for the vast area around including the downtown of cities. Those farmers mostly have high income levels; the need to use straw and other residues as a burning fuel or food for cattle and compost for fertilizer is very low so the rate of straw and other residues burning in the field can reach 60–90%. Their smoke emissions are also realized to be responsible for a number of respiratory illnesses, especially on hot summer days (Ministry of Natural Resources and Environment of Vietnam 2017b). If having a good initial collection and treatment, biomass from agricultural residues can be used as a source of raw materials to produce biomass energy. This type of fuel could become a substitute for fossil fuels used in industries including power generation. In addition, the reduction of fossil fuel consumption also heps to reduce the environmental pollution of land, water, and air, especially the residential and urban areas, which are located near to the industrial parks and power plants. Lately, a lot of countries, including Vietnam, are setting plans to build intelligent communities and cities based on two vital prerequisites for building up the cities and municipalities that are automation and renewable energy. With proper urban planning and harmonious development of alternative energy sources like RE for fossil fuels and the application of smart grid, urban and residential communities in Vietnam will gradually become secure, sustainable, and have intelligent communities and cities in coming years (SDG No. 11). The application of RE technologies in production in general and biomass energy technology in agricultural production in particular will help to manage and use natural resources efficiently. It also will reduce the loss in the production processes while promote to recycle waste generated by applying RE technologies such as biomass energy technology or collecting methane from landfill sites or processing of solid waste into energy pellets, etc., for generating electricity and heating. Consequently, they will contribute to reduce pollutant emissions to land, water, and air environment, thereby reducing negative impacts on human health (SDG No. 12).

Studies have shown that GHG emissions from human activities, especially fossil fuel consumption, is one of the main causes of global warming, increased sea temperatures on the oceans, and increased acidic in seawater. Those reasons are the main causes of bleaching of coral reefs on Earth, destroying marine ecosystems. It is not to mention the environmental disasters causing by discharging of untreated sewage into the marine environment or oil spills from oil transporting ships, etc., which has occurred over time. It will cause a loss of biodiversity in the oceans, a decline in fisheries resources. Nonetheless, waste reduction is equally important, especially the disposed plastic wastes in the ocean. Being a marine and marine-dependent country, Vietnam considers the marine protection and conservation as one of the most vital objectives; therefore, the development of RE is expected to significantly contribute to Vietnam's efforts to conserve the sea (SDG No. 14).

Last but not least, using RE will also contribute to reducing dependency on forests for firewood demand and reduce uncontrol shifting cultivation and deforestation for cultivation. This will reduce human encroachment on natural forests, helping to preserve the biodiversity of the forest (SDG No. 15).

3   Role of Renewable Energy in Vietnam's Nationally Determined Contribution (NDC)

According to the GHG inventory presented in the Initial Biennial Updated Report (BUR1) of Vietnam in 2014, the main GHG emission sources/carbon sinks in Vietnam are energy, agriculture, LULUCF, and waste. The BAU scenario is based on the medium-term scenario of economic development, energy demand, GDP growth by sector, sectoral structure of GDP, population growth, forest and forest land, livestock, and cultivated area for 2020 and 2030 while using IPCC guidelines (Table 1) (Ministry of Natural Resources and Environment of Vietnam 2014). Vietnam committed to working with the international community to respond to

Table 1 The GHG emissions (MtCO$_2$e) in 2010 and projections for 2020 and 2030

Sector	2010	2020	2030
Energy	141.1	389.2	675.4
Agriculture	88.3	100.8	109.3
Waste	15.4	26.6	48.0
LULUCF	−19.2	−42.5	−45.3
Total	**225.6**	**474.1**	**787.4**

Source: Ministry of Natural Resources and Environment of Vietnam (2014)

Table 2 Target to mitigate GHG emissions by 2030 compared to the BAU scenario

Sector	Unconditional		Conditional	
	Target (%)	GHGs (MtCO$_2$e)	Target (%)	GHGs (MtCO$_2$e)
Energy	4.4	29.46	9.8	65.93
Agriculture	5.8	6.36	41.8	45.78
Waste	8.6	4.16	42.1	20.23
LULUCF[a]	50.05	22.67	145.7	66.0
Total	**8%**	**62.65**	**25%**	**197.94**

Source: Ministry of Natural Resources and Environment of Vietnam (2015)
Note: [a]Increased removals

climate change, which is reflected in the range of national policies and specific actions that have been or are being taken to combat climate change. On September 30, 2015, Vietnam sent the UNFCCC Secretariat "Vietnam's INDC." Vietnam signed the Paris Agreement on climate change (PA) on April 22, 2016 and approved the PA on November 3, 2016. From that, Vietnam's INDC has officially become its NDC (Ministry of Natural Resources and Environment of Vietnam 2017a).

Vietnam has identified GHG emission reduction targets for 2030 compared to the BAU scenario (Table 2), which was developed based on the assumption of economic growth in the absence of existing climate change policies. The BAU starts from 2010 and includes the energy, agriculture, waste, and LULUCF sectors. Vietnam's intended contribution to GHG emission reduction efforts in the period of 2021–2030 is as follows (Ministry of Natural Resources and Environment of Vietnam 2015):

- Unconditional contribution: With domestic resources, by 2030 Vietnam will reduce its GHG emissions by 8% compared to BAU, in which:

  - Emission intensity per unit of GDP will decline by 20% compared to 2010 levels.
  - Forest coverage will increase to the level of 45%.

- Conditional contribution: The abovementioned 8% contribution could be increased to 25% if international support is received through bilateral and multilateral cooperation, as well as through the implementation of new mechanisms under the global climate agreement, in which emission intensity per unit of GDP will be reduced by 30% compared to 2010 levels.

For the voluntary contribution (unconditional), energy sector has a fall of only 4.4% and more than doubles (9.4%) with international support (conditional) in target but accounting for nearly half of all GHG emission reductions in total voluntary emission reduction commitments and still account for nearly a third of the total GHG mitigation in the conditional target.

It also shows that the energy sector contributes a great deal to the current and future GHG emission pictures of Vietnam (according to BAU scenario), especially when we study the change in contribution rates with other sectors. For example, GHG reduction target having international support in agriculture sector will be increased more than seven times (from 6.36 to 45.78 $MtCO_2e$) in comparison with unconditional target; for waste sector, having international support will increase almost five times (from 4.16 to 20.23 $MtCO_2e$) in comparison with unconditional target. In short, it can be seen that the GHG reductions in energy sector, whether voluntary or with international support, are the highest in the four major GHG emissions/sinks in Vietnam.

The GHG mitigation options were developed based on the BAU scenario, assuming that new policies will be developed to support the application of mitigation technologies, including energy-saving and the deployment of renewable energy.

Seventeen mitigation options were identified: four on energy-saving and RE in households, two on energy-saving and RE in industry, three on energy-saving and RE in transport, one on energy-saving and RE in commercial services, and seven on energy-saving and RE for electricity production, as shown in Table 3. The options were assessed based on the current state of technology and its application, as well as the objectives set out in sectoral development strategies, such as the national energy development strategy, the transportation development strategy, and the power development plan VII (Ministry of Natural Resources and Environment of Vietnam 2015).

There are eight mitigation options which are RE-based: options E4, E7, E11, E12, E13, E14, E15, and E17. They cover five RE sources including solar, wind power, hydropower, biomass, and biofuel (with ethanol and biogas) and account for about 53% of total mitigation potential in energy sector (Fig. 3). However, not all mitigation options mentioned for the sector will be developed in the near future due to technological and financial barriers, especially with the options chosen to meet the unconditional target. There are just three RE-based mitigation options; options E12, E4, and E13 were selected for the development using domestic sources in a total of 13 selected mitigation options. Though, they account for about 41% of the total mitigation potential in energy sector (equivalent to a reduction of 103 $MtCO_2e$). Those options represent three RE technologies: hydro, solar, and wind because they

Table 3 Mitigation potential of mitigation options in energy sector

Option	Mitigation potential for the entire period (MtCO$_2$e)
E1. High-efficiency residential air conditioning	12.4
E2. High-efficiency residential refrigerators	12.4
E3. High-efficiency residential lighting	38.3
E4. Solar water heaters	16.6
E5. Cement-making technology improvements	16.6
E6. Brick-making technology improvements	19.0
E7. Substitution of ethanol for gasoline in transport	14.2
E8. Passenger transport mode shift from private to public	9.9
E9. Freight transport switch from road	26.7
E10. High-efficiency commercial air conditioning	11.1
E11. Biomass power plants	50.3
E12. Small hydropower plants	83.7
E13. Wind power plants by domestic funding	2.7
E14. Wind power plants by international support	71.8
E15. Biogas power plants	4.4
E16. Ultra-supercritical coal power plants	79.8
E17. Solar PV power plants	12.3

Source: Ministry of Natural Resources and Environment of Vietnam (2015)

Fig. 3 The shares of renewable energies-based options in total mitigation potential in energy sector under Vietnam's NDC

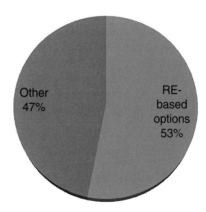

are considered to be easily deployed for the low investment rate, past implementation in Vietnam, and alignment with the relevant sector development plan for the period of 2021–2030. For those reasons, RE are playing the most important role on GHG mitigation potential in energy sector in particular and Vietnam's overall emissions reduction efforts under NDC in the coming time.

4 Challenges of RE Development in Vietnam

Vietnam has not yet made any detailed assessment of the socioeconomic and environmental impacts of all RE sources, and the importance of the link between energy-environment and clean development has not been highlighted. The benefits of developing RE in relation to environmental protection, reducing GHG emissions, adapting to climate change, and contributing to economic growth have not yet been totally reviewed in Vietnam. Therefore, analyses of the socioeconomic and environmental co-benefits of RE used in this chapter are mainly based on visible benefits of these technologies.

Although there are several legal policies on the development of RE in Vietnam, these policies are unsystematic, insufficient, and less supportive for promoting such development (Ngan and Huong 2016). Therefore, it is necessary to develop a strong overall policy for mobilizing and encouraging the involvement of domestic resources, especially the private sector on RE development.

Despite some significant improvement, the overall technology infrastructure for RE development in Vietnam is still lacking. The RE project owners have to import equipment from other countries. The lack of initiative in terms of technology has made the investment cost of building RE projects in Vietnam still high. Therefore, it can be said that finance is still one of the biggest barriers to the development of RE in Vietnam. However, some studies pointed out that developing countries should invest more on renewable energies and implement the RE policy as early as possible to avoid a carbon-intensive lock-in that makes the 2050 target set under Vietnam's RE development strategy too expensive to reach (Tran et al. 2016).

It seems that individual policies of cities and localities are often more ambitious than national ambitions (Frankfurt School-UNEP Centre/Bloomberg New Energy Finance 2017). This is understandable because in practice, the policies and plans of localities are based on the specific characteristics of RE exploitation in each region and the development priorities of each locality and city. In the short term, they should be the main and most realistic drivers for the development of RE in Vietnam.

5 Conclusions

In the context of fossil fuel sources that have been seriously impacting the environment and are becoming to deplete as well as no more reliable alternative energy technology to be found, RE development is an unavoidable roadmap that, sooner or later, countries including Vietnam have to go. Although RE technologies have not yet fully researched and applied, RE sources will play a very important role in Vietnam's socioeconomic development, minimizing adverse impacts on the environment.

The RE accounts for more than half the potential for reducing GHG emissions in the energy sector that Vietnam has committed under the NDC. However, to promote these potentials, the international support will play a very important role.

In order to achieve the goals of the 2030 Agenda for Sustainable Development, Vietnam will need to take really strong actions in all aspects, in which RE is considered to be a "multi-target arrow" for Vietnam based on its important contributions both direct and indirect to meet the setting goals for sustainable development under Vietnam's National Action Plan to implement the 2030 Agenda for Sustainable Development.

References

Cuong ND (2012) An overview of current status and trends of Vietnam's renewable energy market. Access date: 25 Sept 2017

Department for Business, Energy & Industrial Strategy (2016) UK greenhouse gas emissions quarterly official statistics: Q2 2016. Crown copyright, London

EVN (2017) Annual report 2016. Vietnam Electricity, Hanoi

Frankfurt School-UNEP Centre/Bloomberg New Energy Finance (2017) Global trends in renewable energy investment 2017

General Statistics Office (2016) Vietnam socio-economic report 2015. http://vea.gov.vn/vn/truyenthong/tapchimt/nctd42009/Pages/Ph%C3%A1t-tri%E1%BB%83n-n%C4%83ng-l%C6%B0%E1%BB%A3ng-s%E1%BA%A1ch-g%C3%B3p-ph%E1%BA%A7n-gi%E1%BA%A3m-thi%E1%BB%83u-%C3%B4-nhi%E1%BB%85m-m%C3%B4i-tr%C6%B0%E1%BB%9Dng.aspx. http://nangluongVietnam.vn/news/vn/dien-hat-nhan-nang-luong-tai-tao/thuc-trang-nang-luong-tai-tao-viet-nam-va-huong-phat-trien-ben-vung-(ky-1).html

Huong G (2012) Clean energy development contributes to reducing environmental pollution. http://vea.gov.vn/vn/truyenthong/tapchimt/nctd42009/Pages/Ph%C3%A1t-tri%E1%BB%83n-n%C4%83ng-l%C6%B0%E1%BB%A3ng-s%E1%BA%A1ch-g%C3%B3p-ph%E1%BA%A7n-gi%E1%BA%A3m-thi%E1%BB%83u-%C3%B4-nhi%E1%BB%85m-m%C3%B4i-tr%C6%B0%E1%BB%9Dng.aspx. Access date 21 Sept 2017

Huong HTT (2014) The current situation of energy in Vietnam and the direction of sustainable development. Access date: 5 Sept 2017

International Energy Agency (2016) Decoupling of global emissions and economic growth confirmed https://www.iea.org/newsroom/news/2016/march/decoupling-of-global-emissions-and-economic-growth-confirmed.html. Access date: 21 Sept 2017

Ministry of Natural Resources and Environment of Vietnam (2014) Initial biennial updated report of Vietnam to the UNFCCC

Ministry of Natural Resources and Environment of Vietnam (2015) Technical report of Vietnam's intended nationally determined contribution

Ministry of Natural Resources and Environment of Vietnam (2017a) Second biennial updated report of Vietnam to the UNFCCC

Ministry of Natural Resources and Environment of Vietnam (2017b) National status of environment report: urban environment. http://www.ievn.com.vn/tin-tuc/Tong-quan-ve-hien-trang-va-xu-huong-cua-thi-truong-nang-luong-tai-tao-cua-Viet-Nam-5-999.aspx

Ngan NVC, Huong NL (2016) Vietnam's renewable energy – an overview of current status and legal normative documents. Can Tho Univ J Sci, Special issue (Renewable Energy): 92–105

The Prime Minister of Government (2015) Decision no. 2068/QD-TTg: approving the development strategy of renewable energy of Vietnam by 2030 with a vision to 2050, November 25

The Prime Minister of Government (2017a) Decision no. 622/QD-TTg: approving the national action plan to implement the 2030 Agenda for sustainable development, May 10

The Prime Minister of Government (2017b). Decision no. 11/2017/QD-TTg: on mechanism for encouragement on development of solar power in Vietnam, April 11

Tran TT, Fujimori S, Masui T (2016) Realizing the intended nationally determined contribution: the role of renewable energies in Vietnam. Energies 9:587

Vietnam's National Agency for Science and Technology Information (2015) Overview of "potential for renewable energy development in Vietnam"

The Impact of Tenaga Suria Brunei Power Plant on Natural Gas Saving and CO_2 Avoidance

Muhammad Nabih Fakhri Matussin

1 Introduction

Renewable energy deployment is one of the measures to diversify the consumption of the fuel such as oil and natural gas for export, thus reducing dependency in such resources for electricity generation, as well as GHG emissions from power plants. Several countries have been devising a range of policies to encourage widespread use of renewable energy technologies to improve energy security and reduce fossil fuels dependency. For example, ASEAN Member States (AMS) have set a target of achieving 23% of the total primary energy supply by 2025, representing a significant increase of 10% from its 2015 level (ACE 2017). Table 1 illustrates AMS countries' respective renewable energy targets to be achieved.

In the context of Brunei Darussalam, the country's Energy White Paper acknowledges the effect of climate change in the effort to ensure domestic economic and energy securities, despite the country possessing abundant reserves of oil and natural gas. Therefore decarbonisation of the primary energy demand and supply through energy conservation and renewable energy is crucial. The Energy White Paper aims to further expand the renewable energy by generating at least 10% from it into the total electricity generation mix, corresponding to about 954,000 MWh by 2035 (EIDPMO 2014). At present, the renewable energy in Brunei Darussalam constitutes 1.2 MWp of installed capacity, coming from Tenaga Suria Brunei (TSB) solar PV power plant, the 8 kWp grid-connected solar PV system in Kampong Ayer and other small-scale off-grid solar PV systems.

M. N. F. Matussin (✉)
Renewable Energy and Climate Change, Brunei National Energy Research Institute, Bandar Seri Begawan, Brunei

Science and Technology Research Building, Faculty of Science, Universiti Brunei Darussalam, Gadong, Brunei
e-mail: nabih.matussin@bneri.org.bn

© Springer International Publishing AG, part of Springer Nature 2018   57
H.-Y. Chan, K. Sopian (eds.), *Renewable Energy in Developing Countries*,
Green Energy and Technology, https://doi.org/10.1007/978-3-319-89809-4_4

Table 1 AMS countries renewable energy targets

AMS country	Renewable energy target
Brunei Darussalam	954 GWh by 2035
Cambodia	2241 MW by 2020
Indonesia	46,307 MW by 2025
Lao PDR	951 MW by 2025
Malaysia	21,370 MW by 2050
Myanmar	472 MW by 2016
Philippines	15,236 MW by 2030
Singapore	350 MWp by 2020
Thailand	19,684 MW by 2036
Vietnam	45,800 MW by 2030

Source: ACE (2016)

As mentioned earlier, through grid-connected renewable energy deployment, net-exporting countries could significantly save fossil fuels for export. It is estimated that cumulatively around 2.5 billion barrels of oil equivalent could be saved between 2015 and 2030 within the Gulf Cooperation Council countries, which translates to approximately between USD 55 billion and USD 87 billion, subject to oil and gas prices (IRENA 2016). Similarly, the MENA region could see an accumulated saving of 1120 Mtoe between 2017 and 2030, corresponding to avoidance of around 84 and 428 Mt. of CO_2 emissions in 2020 and 2030, respectively (MENARA 2017). In Brunei Darussalam, although the establishment of TSB since 2010 has yielded absolute and monetary savings to the government, there has been no official records or studies being made to quantify these savings. Therefore this chapter looks into quantifying the amount of natural gas and CO_2 saved by TSB generation between 2011 and August 2017 and estimates the future savings up to 2035.

2 Tenaga Suria Brunei Power Plant

The Tenaga Suria Brunei (TSB) is a 1.2 MWp Solar PV Power Generation Plant (Fig. 1) jointly implemented by the Government of Brunei Darussalam and Mitsubishi Corporation of Japan, as part of Mitsubishi's Corporate Social Responsibility (CSR). Situated in an old diesel power plant site in Seria, TSB is capable of providing electricity to about 200 homes. It has a global coordinates of 4.61°N, 114.34°E and altitude of 4.6 m above mean sea level. The TSB was interconnected to the national grid and started operation in May 2010, from which the demonstration phase was done until October 2013. The data collected were analysed to evaluate the performance of six different solar PV modules and to assess which PV panel that performs best under local conditions. The six different solar PV modules include (i) monocrystalline silicon (m-cSi), (ii) polycrystalline silicon (p-cSi), (iii)

Fig. 1 Tenaga Suria Brunei

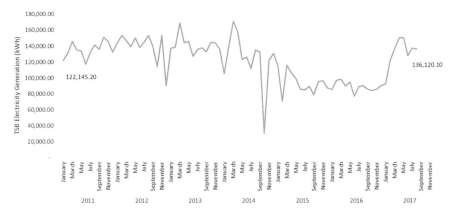

Fig. 2 Electricity generation at TSB (January 2011–August 2017). (Source: Department of Electrical Services 2017)

amorphous silicon (a-Si), (iv) multi-junction silicon (tandem), (v) copper-indium-selenium (CIS) and (vi) heterojunction with intrinsic thin layer (HIT).

As shown in Fig. 2, the generation is highly variable from month to month due to fluctuations in daily solar irradiation. Despite that, there is a gradual downward trend detected between January 2011 and July 2015, from 122 MWh to 85 MWh, representing a decrease of 30%. This could have been driven by the decrease in the efficiency of solar panels, which was mitigated through replacement of new ones. As a result, beyond July 2015, the generation started to pick up which eventually rose to 136 MWh in August 2017. It should be noted that these readings were taken at a uniform tilted angle of 5°. On a yearly basis, the generation decreased by 6% from 2011 to August 2017.

The Brunei National Energy Research Institute (BNERI) has recently carried out a feasibility study to assess the potential expansion of TSB on some of the available nearby areas. By installing polycrystalline silicon PV panels on an additional area of 24 ha, it is envisaged that a minimum of 27 MWp could be added into the existing system (Pacudan 2015).

3   Methodology on Estimating the Savings

The approach is based on the compiled electricity generation data from January 2011 to August 2017, the estimations of natural gas saving, its corresponding monetary savings and CO_2 abatement were made using Microsoft Excel software. The natural gas saving was calculated by taking into account the grid's transmission and distribution losses and the estimated average current heat rate of a single-cycle natural gas-fired power plant in Brunei. Precisely, the electricity generated by TSB can be translated into electricity generated at natural gas-fired power plant before passing through the 66 kV transmission power lines. This saved electricity was then multiplied with the average estimated heat rate of a single-cycle natural gas-fired power plant in Brunei, yielding the natural gas saving. This approach is based on the US EPA[1] (2015) methodology, which calculated fuel and CO_2 savings for combined heat and power (CHP) systems in the USA. In formulas, we have

$$\Delta NG = [(E_{TSB} \times TD_{Loss}) + E_{TSB}] \times h \tag{1}$$

where
ΔNG is the natural gas saving (MMBtu);

E_{TSB} is the electricity production from TSB (kWh);

TD_{Loss} is the average percentage of transmission and distribution losses in the Brunei grid; and.

h is the average estimated heat rate of a single-cycle natural gas power plant in Brunei (at 0.01241 MMBtu/kWh).

The corresponding monthly monetary savings can be easily calculated by multiplying the natural gas saving with monthly LNG import price spot to Japan. In Brunei, 10% of the natural gas produced is being used domestically. The remaining 90% is liquefied and then exported to countries such as Japan and South Korea, with the former accounting for about 85% of LNG export revenues (Hong 2015). In this regard, it can be assumed that the savings are affected by the changes in the import price spot of LNG to Japan.

---

[1]The US Environmental Protection Agency.

Table 2 Transmission and distribution loss estimates

Year	TD_{Loss} (%)
2011	7.168
2012	6.183
2013	9.291
2014–2035	6.414

Source: World Bank (2014)

$$\Delta NG_{Monetary} = \Delta NG \times \varnothing \tag{2}$$

where:

$\Delta NG_{Monetary}$ is the monetary natural gas saving (USD) and

$\varnothing$ is the monthly average LNG import spot price to Japan (USD/MMBtu).

The avoided CO_2 reflects the amount of CO_2 reduced from the atmosphere while producing the same amount from a reference plant, i.e. natural gas power plant in Brunei. This can be easily calculated by multiplying the electricity generation from TSB with grid emission factor, as shown in the following equation:

$$\theta = E_{TSB} \times \varepsilon \tag{3}$$

where

θ is the amount of avoided CO_2 (kg) and

ε is the grid emission factor, at 0.82 $kgCO_2$/kWh (Brander et al. 2011).

Due to the unavailability of the exact transmission loss value, the TD_{Loss} value was used as a satisfactory approximate. The World Bank (2014) provided the estimates of this value for years between 2010 and 2014, as shown in the Table 2. For years 2015 until 2035, the author assumed the 2014 value throughout.

Based on Fig. 3, generally the price per MMBtu of LNG fell gradually from USD 11.45 in January 2011 to USD 8.30 in August 2017, representing a percentage decrease of 28% (YCharts 2017), owing to decrease in Japanese demand for LNG. It is envisaged that the price in 2018 will further drop to USD 7.70, from which it will be slightly increased to USD 8.20 in 2035 (World Bank 2017), as shown in Fig. 4.

The same methodology was also applied for forecasting future savings. PVsyst software was used in the feasibility study of 27 MW system expansion at TSB. Typical polycrystalline solar PV modules and inverter models available in the market were employed in the software. These panels in the simulation were inclined at 5°, consistent with the panels' inclination in the existing TSB. Yearly degradation of panels of 1% was used and an average of 98% availability based on average inverter manufacturer's guarantee (Pacudan 2016). The author assumed that the current technology and capacity of 1.2 MW at TSB is to remain unchanged until 2020, while construction works on addition of 27 MW system would be ongoing. It is also assumed that Mitsubishi would fund for this project as part of their CSR. Thereafter, by early 2020 it is envisaged that the new 28.2 MW system would be able to start feeding the electricity into the national grid.

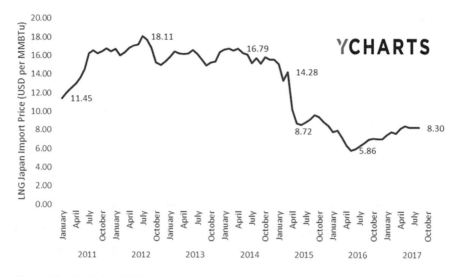

Fig. 3 Monthly Japan LNG import price. (Source: YCharts 2017)

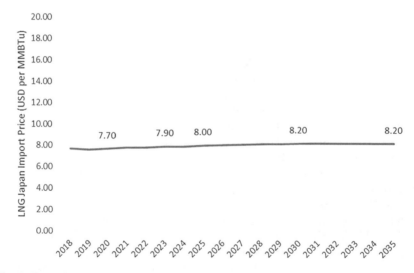

Fig. 4 Yearly forecast Japan LNG import price. (Source: World Bank 2017)

4 Results

4.1 Historical Savings

The monthly savings were significantly affected by fluctuations in electricity generation and LNG import price (Fig. 5). The natural gas savings grew from 1625 MMBtu to 1798 MMBtu between January 2011 and August 2017, an increase of

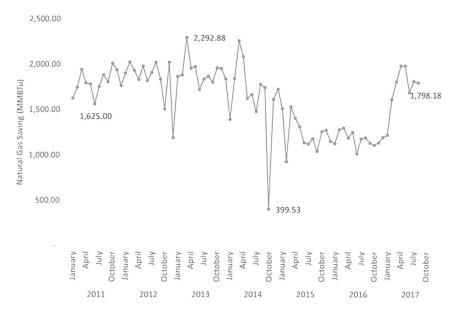

Fig. 5 Monthly natural gas saving

10.7%. Within that period, March 2013 had the highest saving, amounting to about 2292 MMBtu, while the lowest saving of only 399 MMBtu was recorded in October 2014. However, the corresponding monetary savings decreased by 19.8%, from USD 18,606 in January 2011 to USD 14,924 in August 2017, with the largest saving occurred in March 2014, amounting to USD 37,369, and the lowest in October 2014, corresponding to about USD 6348 (Fig. 6). Cumulatively, the TSB has generated savings of around 128,964 MMBtu of natural gas as of August 2017 from 1625 MMBtu in January 2011 (Fig. 7). This translates to cumulative savings of USD 1.730 million from USD 0.0186 million. Annually, the natural gas and its monetary savings decreased by 6.1% and 13.7%, respectively.

The generation of electricity from TSB has also reduced the overall CO_2 emission from conventional power plants in Brunei Darussalam. Based on the estimated national grid emission factor of 0.82 kg CO_2 per kWh, the CO_2 avoidance ranged from 100,159 to 111,618 kg between January 2011 and August 2017 (Fig. 8). The highest recorded was in March 2014, with a reading of 140,158 kg, while the lowest was in October 2014, with a reading of 24,799 kg. Cumulatively, the avoidance increased at a rate of 5.6%, from 100,159 to 7.961,298 kg as of August 2017 (Fig. 9).

Fig. 6 Monthly monetary natural gas saving

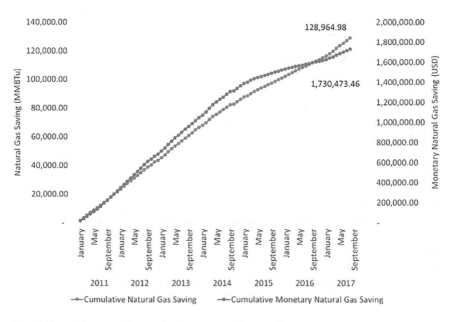

Fig. 7 Cumulative natural gas and monetary natural gas savings

Fig. 8 Monthly CO_2 avoidance

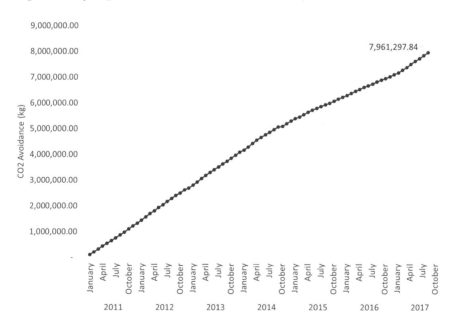

Fig. 9 Cumulative CO_2 avoidance

4.2 Projected Savings

Electricity generation at TSB is assumed to continue decreasing annually by 6% in 2018 and 2019, following similar trend between January 2011 and August 2017. Consequently, the yearly gas saving for 2018 and 2019 amounted to 13,071 MMBtu and 989,794 MMBtu, respectively. In monetary terms, these correspond to USD 100,651 and USD 93,383 in 2018 and 2019. The corresponding CO_2 avoidance totalled 811,395 kg and 762,711 kg, respectively.

Based on the modelling on the PVsyst by Pacudan (2016), the first year operation of 27 MW system at TSB would generate about 41,928 MWh of electricity. Combining this with the electricity yield of the existing 1.2 MW system of about 874,327 kWh, cumulatively the total electricity generation would amount to 42,802 MWh in 2020. This represents an increase of 41,872 MWh, corresponding to a percentage increase of 4502% relative to generation in 2019. Beginning 2020, at a yearly degradation rate of 1%, electricity generation would decrease to 40,704 MWh in 2025, 38,709 MWh in 2030 and eventually reaching 36,812 MWh in 2035, as shown in Fig. 10.

The increase of electricity generation between 2019 and 2020 would be able to displace about 553,140 MMBtu more of natural gas, from just 12,287 MMBtu to 565,427 MMBtu (Fig. 11), so would the corresponding monetary saving which increased from USD 93,383 to USD 4.353 million (Fig. 12). Between 2020 and 2035, the gas saving would decrease from 565,427 MMBtu to 486,301 MMBtu, at a rate of 1% per year. Monetary gas saving would also experience reduction from USD

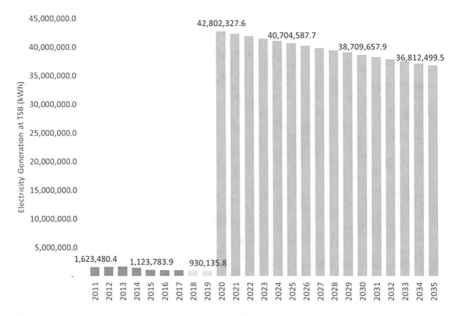

Fig. 10 Historical and projected annual TSB electricity generation

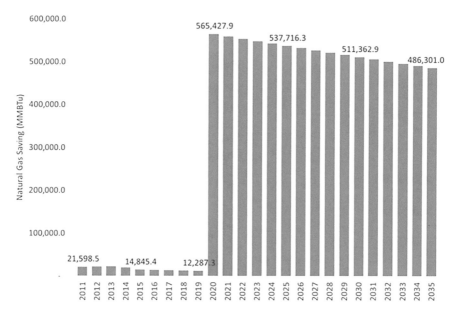

Fig. 11 Historical and projected annual natural gas savings

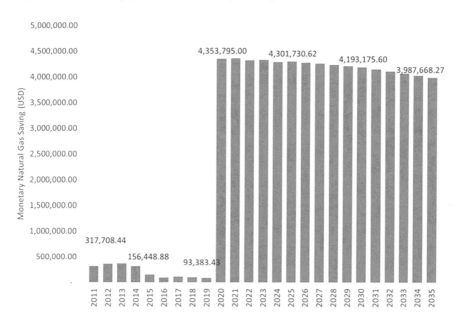

Fig. 12 Historical and projected annual monetary natural gas saving

4.353 million in 2020 to USD 3.987 million in 2035, at a rate of 0.5% per year. The lower rate than that found in electricity generation and natural gas saving is due to the trend in the forecast of LNG price, which experiences gradual increase from USD 7.70 per MMBtu in 2020 to USD 8.20 per MMBtu, as shown in Fig. 4 earlier.

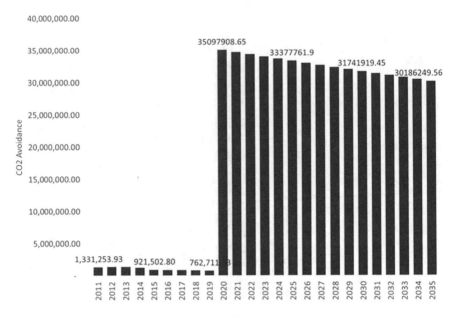

Fig. 13 Historical and projected annual CO_2 avoidance

Subsequently the electricity generation from the TSB expansion would potentially further reduce about 34,335,197 kg (or 34,335 tonnes) of CO_2 emission between 2019 and 2020. Between 2020 and 2035, the CO_2 avoidance would also decrease from 35,097,908 kg (35,097 tonnes) to 30,186,249 kg (30.186 tonnes), at a rate of 1% (Fig. 13). Cumulatively, the TSB expansion project would be able to save the government about 8.553 million MMBtu of natural gas, corresponding to about USD 69.421 million (Fig. 14). This would also help avoid cumulatively around 530,887 tonnes of CO_2 by the year 2035 (Fig. 15).

5 Conclusion

Savings on natural gas, its monetary value and the CO_2 avoidance were calculated using methodology adopted by the US EPA. In a nutshell, any renewable energy project connected to the grid could provide significant natural gas saving for Brunei Darussalam, since the country runs on natural gas for electricity generation. Reduction in the use of natural gas for electricity generation could subsequently help contribute to the reduction of CO_2 emissions. The monetary saving achieved could be used to maintain and expand the current TSB system or to fund for other renewable energy projects that are vital for the country to reach its 10% target in renewable energy mix by 2035.

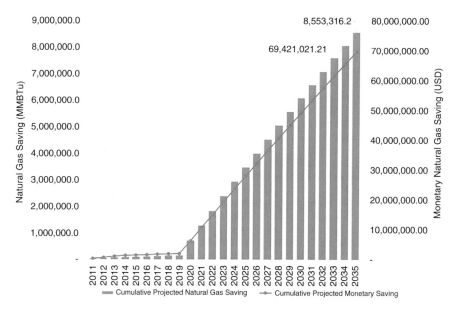

Fig. 14 Cumulative historical and projected natural gas and monetary savings

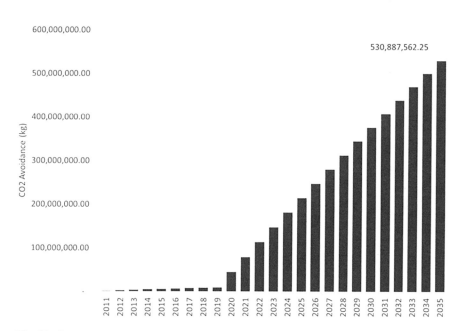

Fig. 15 Cumulative historical and projected CO_2 avoidance

References

ASEAN Centre for Energy (ACE) (2016) The current status of RE and its target in ASEAN member states. September 2016

ASEAN Centre for Energy (ACE) (2017), Development of Renewable Energy Outlook for ASEAN – a REmap 2030 Analysis.

Brander M, Sood A, Wylie C, Haughton A, Lovell J (2011) Electricity-specific emission factors for grid electricity, Technical Paper, Ecometrica, August 2011

Department of Electrical Services (2017) Tenaga Suria Brunei electricity generation (raw data), August 2017

Energy and Industry Department, Prime Minister's Office (2014) Energy white paper

Hong Z (2015) China and ASEAN: energy security, cooperation and competition. ISEAS – Yusof Ishak Institute, p 94

IRENA (2016) Renewable energy market analysis – the GCC region

Middle East and North Africa Regional Architecture (MENARA) (2017) How can renewable energy help contribute to the development of the MENA countries? Future notes, July 2017

Pacudan R (2015) 27MWp Tenaga Suria Brunei solar PV expansion project: yield analysis and levelized cost analysis. Brunei National Energy Research Institute, Brunei Darussalam

Pacudan R (2016) Implications of applying solar industry best practice resource estimation on project financing, Brunei National Energy Research Institute, Brunei Darussalam, Elsevier. Energy Policy 95:489–497

The World Bank (2014) Electric power transmission and distribution losses (% of output). Website: https://data.worldbank.org/indicator/EG.ELC.LOSS.ZS?locations=BN. Last accessed: 4 Oct 2017

The World Bank (2017) World Bank commodities price forecast. 26 April 2017

US Environmental Protection Agency (2015) Fuel and carbon dioxide emissions savings calculation methodology for combined heat and power systems. Combined heat and power partnership, February 2015

YCharts (2017) Japan liquefied natural gas import price. Website: https://ycharts.com/indicators/japan_liquefied_natural_gas_import_price. Last accessed: 8 Oct 2017

Part II
Technology Development and the Feasibility

The Needs of Solar Energy Technology from the Perspective of Aboriginal People in Tasik Chini

Suhaila Abdul Hamid, Hoy-Yen Chan, Ah Choy Er, Wong Chin Yew, and Kamaruzzaman Sopian

1 Introduction

Orang Asli is a Malay term interpreted as 'original people' or 'first people' (Masron et al. 2013). The term was used by anthropologists and administrators for the 18 sub-ethnic groups generally classified for official purposes as listed in Table 1. They are the indigenous people and the oldest inhabitants of Peninsular Malaysia which live in since about 5000 years ago. Based on the animistic belief inherited by their ancestors, this tribe prefer to live in isolated area such as in the forest, hillside, lakeside or an island, connecting themselves with the nature. It is believed that by connecting to the nature, they would have longer life and lower mortality rate (Shamsuddin 2012). However, these locations are a cold spot of modernisation where electricity is inaccessible, and basic facilities are incomplete. Electrical energy has become an essential of a modern society for lighting, space heating and cooling as well as an energy source for information technology and communication. By supplying this community with modern energy sources such as electricity could be one of the ways to improve their quality of life.

In order to supply the electrical energy in a remote area, solar energy (SE) is a perfect choice to provide the off-grid electricity through the stand-alone system. SE technology commonly known as photovoltaic (PV) system converts solar energy to electricity. It is one of the renewable energy (RE) sources that harnesses sunlight, and it is well known as a promising carbon-free energy for the future (Othman et al. 2013; Othman et al. 2016). PV system is made of an array of semiconductor solar

S. A. Hamid · H.-Y. Chan (✉) · K. Sopian
Solar Energy Research Institute (SERI), Universiti Kebangsaan Malaysia, Selangor, Malaysia
e-mail: hoyyen.chan@ukm.edu.my

A. C. Er · W. C. Yew
Faculty of Social Sciences and Humanities, Universiti Kebangsaan Malaysia, Selangor, Malaysia

© Springer International Publishing AG, part of Springer Nature 2018
H.-Y. Chan, K. Sopian (eds.), *Renewable Energy in Developing Countries*,
Green Energy and Technology, https://doi.org/10.1007/978-3-319-89809-4_5

Table 1 The list of sub-ethnics of Orang Asli in Malaysia

Senoi	Negrito	Melayu Proto
1. Temiar	1. Kensiu	1. Kuala
2. Semai	2. Kintak	2. Kanaq
3. Semoq Beri	3. Lanoh	3. Seletar
4. Jahut	4. Jahai	4. Jakun
5. Mah Meri	5. Mendriq	5. Semelai
6. Che Wong	6. Bateq	6. Temuan

Data source: JAKOA (2017)

cells. This solid-state electronic device converts radiant energy directly into electrical energy by employing the photoelectric effect (Hamid et al. 2014). Although the energy from the sun can only be harnessed during the day, the electricity can be stored in a battery. The system is easy to be installed. The advantages of PV applications in a rural area are as follows (GENI 2014):

1. Environmental-friendly technology
2. Low maintenance
3. Low operational cost
4. Long lifespan
5. Highly portable
6. Ideal for light-emitting diode (LED) lightning
7. Simple installation

The efforts to empower the Orang Asli in terms of socio-economic development have been taking since the Seventh Malaysia Plan (1996–2000). Through the Department of Orang Asli Development (JAKOA) and non-government organisation (NGO) supports, many development and environmental projects have been implemented to improve their socio-economic conditions, upgrade their land infrastructure and enhance their standard of living in terms of education and health.

Through the socio-economic development projects by JAKOA, 68% of the Orang Asli settlements have received electric supply, while 71% received water supply (Alam et al. 2013). The electricity supply includes the on-grid and off-grid electricity to the settlements, whether genset or renewable energy, while the water supply includes pipe to the land and pump to the lakeside settlements. The infrastructure facilities such as roads, bridges and community hall were also upgraded and built. Basic health check-ups run by the County Health Department were done periodically. This is to ensure villagers are equipped with general health knowledge and practice. NGOs have also participated in this community project whether through corporate social responsibility (CSR) or personal contributions. These include supply small solar kit to each family, food supplies, and education fund as well as school aid.

Nonetheless, it is found that many of the solar panels, gensets and pumps installed in the Orang Asli villages were broken and abandoned after the project period. The main factor contributed to this problem is because the Orang Asli community is lack of knowledge on the maintenance and repairing the breakdown equipment. These facilities are very important in their daily activities. If a training session on the equipment maintenance is provided to the local community, continuous electricity and clean water supplies will be able to sustain. For example, the operational of PV stand-alone system and its maintenance are simple and economically feasible. With the knowledge transfer to the Orang Asli community, the provided facilities would be self-sustain within the community.

One of the objectives of this study is to investigate the electricity expenditure of the Orang Asli community at the Tasik Chini in relation to their monthly income. This is important to understand their electricity affordability and analyse the burden of energy expenses over the monthly income. Furthermore, the study also intends to have an overall perspective on the acceptance level of the community towards solar energy technology. In addition, this research is also a community project of the Universiti Kebangsaan Malaysia (UKM), where the selected villages will be installed with solar panels. Prior to the installation, the research team engaged the Orang Asli community to promote the application of solar energy, explaining how this technology will help to reduce their current energy expenditure burden, and to promote their ownership of the facility and hence the commitment of sustaining the facility. This chapter starts by introducing the profile of Orang Asli settlements at the Tasik Chini, followed by research methodology descriptions, research results and conclusions of the findings.

2 Background of Orang Asli at the Tasik Chini

Tasik Chini is a lakeside located at Pekan District, the south-eastern state of Pahang. It is located 231 km away from Kuala Lumpur. Tasik Chini is the second largest freshwater lake in Malaysia after Tasik Bera, which covers an area of 202 ha. The lake is surrounded by a chain of 12 small lakes, which the local people call the lake as "laut" in Malay (sea in English) (Ahmad et al. 2009). The ecosystem of Tasik Chini is rich with various species of freshwater fish and particularly a good breeding place for Sungai Pahang fish (van der Halm 2000). Due to its diverse range of ecosystem and highly potential in the tourism sector, Tasik Chini has been endorsed as the pioneer of Reserve Biosphere in Malaysia (Habibah et al. 2010). This has also been recognised by the United Nations Educational, Scientific, and Cultural Organization (UNESCO) in 2009 in preserving the threaten ecosystems and the sustainability of the livelihood of the community.

There were originally six Orang Asli settlements in the Tasik Chini. Nevertheless, one of the settlements has only three families and later was merged with the other three settlements. In general, the settlements of the tribes can be grouped by land and lakeside area. The land area consists of two villages, which are Kg. Gumum Dalam (main village) and Kg. Ulu Gumum, whereas the lakeside areas are Kg. Tg. Puput, Kg. Cendahan and Kg. Melai. Each of these is led by a chief known as Tok Batin, a respected village elder. The villages of the land area can easily be reached by land transport, while the lakeside area can be reached by boat in 15 min time.

3 Methodology

The settlements of Orang Asli at Tasik Chini were also one of the selected sites by the Orang Asli Department (JAKAO) for Orang Asli development projects. These projects are such as housing, formal education and socio-economic development programmes.

The information collected by this research was based on primary and secondary data. The primary data were collected using a qualitative method from fieldwork activities. The chief and some of the villagers were randomly chosen by the research team to carry out an interview session. A set of a structured questionnaire was prepared, and the conversations were recorded during the interview sessions. The questionnaire is attached in Appendix A. In addition, the interview results were also supported by the field observation by the researchers. On the other hand, the secondary data were the information that obtained from the JAKOA and relevant agencies and other published literatures.

The respondents for this survey are the chief of the tribes (Tok Batin), chief of the villagers (Ketua Kampung) and some of the villagers. Due to the fieldwork activities were carried out in the daytime, majority of the interview respondents were females. The men were out for work and to earn for living, while children were in the school and only housewives, old folks and babies were at home. Nevertheless, since the villagers are kin to each other that they live together and form a big family, the respondents were able to share their views in a representative context, and information collected from the chief of villagers are especially helpful.

The on-field data collection activities were completed in August 2016. During the fieldworks, the research team was assisted by the researchers from the Tasik Chini Research Center (PPTC), who have a good relationship with the Orang Asli community.

4 Result and Discussion

Respondents from these five villages have given their full excellent cooperation during the interview session. The prepared questionnaire was divided into two sub-divisions. The first section was on the social-economic of the Orang Asli community, while the electricity supply is covered in the second section. In the second section, their understanding of RE and the acceptance level on the application of SE technology were investigated.

4.1 Social-Economic of the Tribes in Tasik Chini

In each Orang Asli settlement, the community is led by the chief, known as Tok Batin by the locals. Tok Batin is the most respected and influential person in a village and assisted by the Ketua Kampung (head of the village). These leaders are responsible for managing their own village, understanding the villagers' needs and being the middle person to link with outsiders. Figure 1 shows a picture of Tok Batin in an interview session with the research team.

The total population of Orang Asli at the Tasik Chini are 481 persons. The socio-demographic of the villages is listed in Table 2. Kampung Gumum Dalam and Kampung Ulu Gumum have higher population numbers among the villages. They

Fig. 1 Researchers are interviewing Tok Batin Awang Alok at Kg. Gumum Dalam

Table 2 Socio-demographic of the community

Name of villages	Kg. Gumum Dalam	Kg. Ulu Gumum	Kg. Tanjung Puput	Kg. Cendahan	Kg. Melai
No. of houses	40	32	9	5	3
No. of family	40	53	14	5	3
Adults	100	106	28	10	6
Kids	100	94	10	15	12
Religion	Islam, Christian	Islam, Christian, Animism	Animism	Animism	Animism

are able to access to more resources, including the financial aides from the government and non-government agencies (NGO). They are more explored to the modern living style, equipped with better facilities and so attract more young generation and tribes to move to these two villages. For the other three villages, they are relatively poorer, and some families are sharing houses, especially those who have family bonds, e.g. siblings.

In terms of socio-economic background, the monthly income of land villagers of Kampung Gumum Dalam and Kampung Ulu Gumum is about RM650 per household on an average. Whereas, the other three lakeside villagers can only earn up to RM300 per household on an average. This huge different value of income is because the job opportunities of the lakeside villagers are limited. They used to hunt in the forest, catch fishes at the lake, work in logging industry and perform a cultural performance to the tourists at the village. However, in order to preserve the biosphere and the ecosystem of the forest, the government had prohibited the hunting and logging activities. A number of villagers who work as fishermen are also decreasing due to the population of fish that has been declining caused by illegal loggings and the massive flood in 2012. The cultural performance is no longer active because the number of visitors is decreasing drastically.

To overcome this limitation, the lakeside villagers took initiatives by harvesting forest products and handmaking souvenirs to sell the tourists at the land villages and souvenir shops at the town. Figure 2 shows one of the villagers at Kampung Tanjung Puput making a cane from a stick. Moreover, the villagers also grow their own plants such as herbs, vegetables, cassava and tapioca as their own food supplies.

Fig. 2 Villager is handcrafting the stick to make a cane at Kg. Tg. Puput

Most of the villagers at the land settlement work at the rubber estate and oil palm plantation that are located in the Kampung Gumum Dalam, which are managed by a government agency – Rubber Industry Smallholders Development Authority (RISDA). Some of the villagers who have own vehicles such as motorcycles and cars, they are able to travel and work in the town.

4.2 Electricity Supply

The interviews were focused on the importance of electricity to the Tasik Chini community and also to investigate whether the source of electricity is a factor that the tribes would take into consideration for their daily life. Kampung Gumum Dalam and Kampung Ulu Gumum are able to access the on-grid electricity supply, clean water treatment, stable telecommunication and Internet services since the past 10 years. Their electricity bills cost about RM50 per month per household on an average, which is about 7.69% of their monthly income. Thus they can still afford to pay the bill. The villagers are grateful for the modern facilities and services that they have been provided and did not show interest in solar energy.

On the other hand, villagers in Kampung Chendahan, Kampung Tanjung Puput and Kampung Melai are relatively living in lower quality of life. Due to their geographically remote area, the grid transmission is not able to reach these villages, and thus the electricity was powered by generators. The generators were shared within individual village. The use of electricity is mainly for lighting and space cooling. The villages need to share the cost of 180 l of petrol, which is about RM330 per month (depending on the current price of oil) in fuelling up the generator. For instance, if one generator is to be shared by two families, each family has to pay RM165 for the petrol, and this amount is equivalent to 55% of their monthly household. Whereas, poor families that cannot afford to pay the generator fuel will light up at night by using kerosene lamps. In addition, there are no water treatment facilities in these three villages, and they use and drink water directly from the hill or the lake.

Nonetheless, among all the villagers of these three villages, the chief of Kampung Chendahan is able to earn about RM1200 per month. He has multiple jobs, such as boat driver for the school children and tourists, fisherman and also harvests forest products for sale. He affords to buy a flat television screen and subscribe television channels through Malaysia satellite service provider. This reflects that the electricity usage is linked to income level. In this case, affordable electricity enables his family to live in relatively more "luxurious" lifestyle.

Table 3 shows the perspectives of villagers on the needs of electricity and the willingness to learn on SE. The interviews found that villages with accessible electricity could benefit them to live in better quality of life. Though the villagers are aware of the environmental benefits of SE, only the villagers from the three remote villages are willing to learn about SE. In addition, the remote villagers appreciate the SE more because they have witnessed the technology through the

Table 3 The needs of electricity and solar energy in the Tasik Chini villages

Name of villages	Kg. Gumum Dalam	Kg. Gumum Ulu	Kg. Tg. Puput	Kg. Cendahan	Kg. Melai
1. Level of electricity need	Much needed	Much needed	Much needed	Much needed	Much needed
2. The needs of solar energy	Not needed	Not needed	Much needed	Much needed	Much needed
3. Willingness to learn about solar energy	No	No	Yes	Yes	Yes
4. Current power source	On-grid supply	On-grid supply	Genset and small solar panel	Genset and small solar panel	Genset and small solar panel

solar panel and small solar kit provided by the JAKOA and NGOs, respectively. Nonetheless, knowledge on maintenance and operation needs to be provided to minimise the technology breakdown. For the Kampung Gumum Dalam and Kampung Ulu Gumum, the villagers have been using the grid-connected electricity, and they never had an experience of solar-powered electricity; with such sufficient supply condition, solar energy is thus not their priority.

5   Conclusion

Orang Asli community has a special relation with the nature. They believe that one of their main roles is to protect the natural resources. Hence, it is important to ensure the Orang Asli community has access to the basic facilities while continuously protecting the natural environment.

Based on our findings, the Orang Asli settlements can be grouped into two – land and lakeside settlements. The former are the Kampung Gumum Dalam and Kampung Ulu Gumum, whereas the latter are Kampung Chendahan, Kampung Tanjung Puput and Kampung Melai. Due to the land settlements that are geographically easier to be reached, villagers live in moderate quality of life where the basic facilities and services are provided by the government. The frequent aids sent by individuals, NGOs or government are also relatively more compared to the lakeside settlements. Having sufficient electricity supply, the land settlements showed lower interest in solar energy compared to the lakeside settlement communities.

As the villagers of the lakeside settlements earn averagely less than RM300 per household per month, they live in poorer condition compared to the land settlements, which the average income is about RM650 per month. Furthermore, because of inaccessible grid transmission, the electricity source is from generators. About 55% of their incomes contribute to purchase the gasoline is a burden to the community. Nonetheless, living in such limited modern services conditions, they always practice and converse the traditional Orang Asli living styles that integrated with the natural environment, which they believe this is the main reason of having lower mortality rate compared to the land settlements.

Overall, solar energy is highly accepted by the Orang Asli community because of its ability to provide electricity without compromising the natural environment. It is also a good solution for remote area electrification. However, the interest to learn about solar energy is only 16.84% of the total population of the communities, who are from the lakeside settlements. The villagers are aware that in the long term, solar energy not only can improve their quality of life but also will reduce their living expenses on fuel consumption for the generators. On the other hand, Orang Asli community who has access to the grid has shown no interest to learn about this technology.

The willingness to learn SE is closely related to the electricity affordability which can be seen here that the poorer villages appreciate more the SE technology. Furthermore, knowledge transfer is a key factor in order to sustain the SE technology, to minimise the equipment breakdown and to provide the basic knowledge on minor reparation.

Acknowledgements The authors would like to thank the Universiti Kebangsaan Malaysia (UKM) for research grants of KOMUNITI-2014-011 and AP-2014-022. The research team appreciates the assistance and support given by the Tasik Chini Research Center (PPTC), UKM, throughout this study.

Appendix A: Questionnaire of the Acceptance Level of RE to the Orang Asli Community in Tasik Chini

Section 1: Background/socio-economic status

1. Name of Tok Batin (Chief of the tribe):
2. Name of Ketua Kampung (Chief of the village):
3. How many houses in the village:
4. How many families in the house:
5. How many number of the villagers, according to gender and age:
   Adult male: _____ Adult female: _____ Kids/infants: _____
6. What are your religious beliefs and your tribal group:
7. What is your source of food and drink:
8. What is your source of income:
9. Do you receive any aid from the government or non-government organisation. If yes, please state:
10. What are the facilities and infrastructures provided in this village:
11. What are your and other villagers' daily activities:

Section 2: Source of electricity

1. What is the source of electricity:
2. What are the electrical appliances in your house:
3. What is your level of knowledge in terms of repair and maintenance electrical appliances?

    Very poor:_____ Poor: _____ Moderate: _____ Good: _____
    Excellent: _____
4. Do you receive any contribution in terms of electrical appliances from any agencies?

    If yes, please state:
5. What is your level of electricity needs in daily routine:

    Not needed:_____ Need: _____ Much needed: _____.

Section 3: Source of renewable energy → solar energy

1. What is your perception and acceptance level towards new technology:

    • If it is bring benefit, how far it can help to improve your quality of life?
    • If it is bring bad, does it contradict with your original ancestors' beliefs and lifestyle?

2. What do you know about solar energy? Do you know how it works and its application:
3. If you are given a solar panel with a complete equipment including the training, will you accept it? Would you be able to take care the equipment? If yes/no, please state the reason.
4. Are you ready to accept the solar energy technology to be installed in your village?

    • Yes: _____ No: _____

5. Are you willing to learn about solar energy technology?

    • Yes: _____ No: _____

6. To which extent do you think you need such technology:

    • Not needed:_____ Need: _____ Much needed: _____

References

Ahmad H, Jusoh H, Idris M, Kurnia A, Yussof I, Omar M (2009) Strengthening the economy of natural lakes regional development in ECER Chini. Prosiding Perkem 1:180–194

Alam SS, Omar NA, Ahmad MS, Siddiquei HR, Nor SM (2013) Renewable energy in Malaysia: strategies and development. Environ Manag Sust Dev 2(1):51–66

Global Energy Network Institute (GENI) (2014) Sustainable energy solutions for rural areas and application for groundwater extraction. Online: www.geni.org [1.3.2017]

Habibah A, Hamzah J, Mushrifah I (2010) Sustainable livelihood of the community in Tasik Chini biosphere reserve: the local practice. J Sust Dev 3(3):184–196

Hamid SA, Othman MY, Sopian K, Zaidi SH (2014) An overview of photovoltaic thermal combination (PV/T Combi) technology. Renew Sust Energ Rev 38:212–222

Jabatan Kemajuan Orang Asli Malaysia (JAKOA). Suku Kaum/Bangsa. Website: http://www.jakoa.gov.my/orang-asli/info-orang-asli/suku-kaumbangsa/. Accessed date: 15 Sept 2017

Masron T, Masami F, Ismail N (2013) Orang Asli in Peninsular Malaysia: population, spatial distribution and socio-economic condition. J Ritsumeikan Soc Sci Hum 6:75–115

Othman MY, Ibrahim A, Jin GL, Ruslan MH, Sopian K (2013) Photovoltaic-thermal (PV/T) technology – the future energy technology. Renew Energy 49:171–174

Othman, MY, Hamid, SA, Tabook, MAS, Sopian, K, Roslan, MH, Ibrahim, Z (2016). Performance analysis of PV/T Combi with water and air heating system: An experimental study. Renewable Energy 86: 716–722

Shamsuddin AH (2012) Development of renewable energy in Malaysia strategic initiatives for carbon reduction in the power generation sector. Proc Eng 49:384–391

van der Halm PA (2000) History and profile of the Orang Asli at Tasik Chini. Retrieved from https://perswww.kuleuven.be/~u0084530/srigumum/doc/history.html

Feasibility Study of Solar-Powered Hydroponic Fodder Machine in Bangladesh

Md. Tanvir Masud and Sajib Bhowmik

1 Introduction

Farming is a traditional and the most common profession in Bangladesh. Most of the people of Bangladesh live in the villages, and most of the villagers are involved in farming. One of the main reasons of having common interest to farming is because the land is fertile and plain. Another interesting factor is almost all Bangladeshi people are born farmers. However, Bangladesh is among the most vulnerable countries to climate change, which poses a long-term threat to the country's agricultural sector, particularly in areas affected by flooding, saline intrusion, and drought. Faster and more inclusive rural growth with job creation will require greater agricultural diversification together with more robust rural nonfarm enterprise development. A shift in production from rice to higher-value crops will significantly reduce malnutrition, trigger more rapid growth in incomes, and create more and better on-farm and nonfarm jobs, especially for women and youth. Livestock and fisheries will also offer tremendous potential for reducing malnutrition and increasing incomes and job opportunities in a severely land-constrained economy. Nonetheless, the present condition is the lack of adequate government support.

Land crisis is one of the critical challenges for agriculture and power generation in Bangladesh. Bangladesh has 72% access to electricity. So, solar-powered hydroponic fodder machine can play a game-changing role in rural areas. It will produce crops as well as grass for dairy farms without using grid power in rural areas.

Hydroponics is a system of agriculture whereby plants are grown without the use of soil as a media. There are two chief merits of the soilless cultivation of plants. First one is higher crop yields and the second is that it can be used in places wherein ground agriculture or gardening is not possible. This system has several advantages, is cost-effective, and contains all the nutrients the animal requires. No nutrition

Md. T. Masud (✉) · S. Bhowmik (✉)
Sustainable And Renewable Energy Development Authority (SREDA), Dhaka, Bangladesh

© Springer International Publishing AG, part of Springer Nature 2018
H.-Y. Chan, K. Sopian (eds.), *Renewable Energy in Developing Countries*,
Green Energy and Technology, https://doi.org/10.1007/978-3-319-89809-4_6

pollution is released into the environment because of the controlled system. Pests and diseases are easier to get rid of than in soil. Using of solar power will increase the sustainability of the project.

So we have gone through a thorough study about the feasibility of solar-powered hydroponic machines in Bangladesh to find out a probable solution of land crisis, malnutrition problem, and power generation.

2   Livestock Production Scenario of Bangladesh

Livestock is an integral component of the complex farming system in Bangladesh as it is not only a source of meat protein but also a major source of farm power services as well as employment. The livestock sub-sector provides full-time employments for 20% of the total population and part-time employments for another 50%. The poultry meat alone contributes a substantial 37% of the total meat production in Bangladesh (FAO 2005).

The GDP contribution of this sub-sector has been a modest 1.66% annually in 2015–2016 (FAO 2005). However, the sector's actual contribution has been consistently underestimated as the value added in draft power used in farm operation, threshing, sugarcane and oilseed crushing, local transport, dung for cooking fuel, and manure for fertilization of crop fields was not taken into account. An estimate of the uncounted sectoral contribution of livestock indicates a foregone value of three times the amount of official GDP attributed to this sector. According to the DLS survey 2015–2016, share of livestock to agriculture GDP is about 14.21%, and GDP volume is about BDT 329100 million (Hossan 2017). Therefore, livestock is playing an important role in GDP growth of Bangladesh (Huque and Sarker 2014; BBS 2016).

As we can see from Table 1, the livestock population is increasing yearly for each sectors starting from cattle to goat. So it is also required to ensure this increased volume will get the best use in terms of demand and market price for the sustainability of the sector. We need to look into the demand vs supply curve to assess the situation, and Table 2 will help us in this cause.

Table 1 Year-wise livestock population of Bangladesh (00,000)

Livestock species	2011–2012	2012–2013	2013–2014	2014–2015	2015–2016
Cattle	231.95	233.41	234.88	236.36	237.85
Buffalo	14.43	14.50	14.57	14.64	14.71
Sheep	30.82	31.43	32.06	32.70	33.35
Goat	251.16	252.77	254.39	256.02	257.66
Total livestock	528.36	532.11	535.90	539.72	543.57

Source: FAO (2005), Bhattacharjee and Khatun (2016)

Table 2 Animal and protein production and availability

Products	Demand		Production		Availability		Deficiency	
Year	2011	2016	2011	2016	2011	2016	2011	2016
Milk	11.86	14.69	1.78	7.27	37.51	125.59	10.08	7.41
Meat	5.69	7.05	0.78	6.15	16.44	106.21	4.91	0.90

Source: FAO (2005), Bhattacharjee and Khatun (2016)

Table 3 Contributions of livestock contribution in GDP in Bangladesh from 2011 to 2016

	2011–2012	2012–2013	2013–2014	2014–2015	2015–2016
Contribution to GDP (%)	1.90	1.84	1.78	1.73	1.66
Growth rate of GDP (%)	2.68	2.74	2.83	3.10	3.21

Source: FAO (2005), Bhattacharjee and Khatun (2016)

So the above table is a clean indication that the deficiency between demand and supply is decreasing, and Table 3 will show how it is making an impact on the national GDP.

3   Constraints to Livestock Production in Bangladesh

The five major factors being the most important in limiting animal production in Bangladesh are:

- Poor nutrition/availability of animal feed and land on which to grow fodder
- Poor organization of primary producers
- Poor extension of information to producers
- Land ownership and absence of effective land resource use planning
- Market infrastructure

Most cultivable land in Bangladesh is used for growing food crops. There is virtually no land that is exclusively available for fodder production. Feed resources available to ruminants are therefore mostly crop residues (rice straw) supplemented with green fodder and weeds from cultivated fields. Cows in semicommercial farms are stall-fed with a mixture of concentrates with water, ad-lib rice straw throughout the day, and, at night, 2–5 kg of hand-collected green grass. Concentrate mixtures typically consist of pulse bran, wheat bran, rice polish, oil cakes, or broken rice (0.5–1 kg/day). Some farmers feed chopped rice straw soaked in water with concentrates and a small quantity of green grass. Indigenous milking cows owned by landless and marginal farmers, such as the members of the Milk Vita village cooperatives, are fed with small quantities of rice straw, often treated with urea, and other crop residues and green roughage scavenged from crop weeds, grasses, and herbs from bunds, roadsides, and railway embankments, shrub and tree foliage, and aquatic plants and weeds. Cooperative members store rice straw in stacks close to their elevated homesteads to serve as a feed reserve during the annual flood season. About 50% of ruminant forage comes from straw and the balance from these other sources. Crop residues can potentially supply about 30 million tonnes of forage annually. It is probable that only about half is currently fed to stock, the remainder being lost during harvesting or used for fuel or for construction purposes.

Currently, existing feed resources meet the subsistence requirement of livestock. In most cases the diet does not involve any additional expenditure by the farmer as it is made up from his/her own crops, grazing and scavenging on common or waste-land. However, using proven technology to improve, for example, forage storage and digestibility, along with efficient concentrate supplementation, milk (and meat) production can be substantially enhanced as experienced in the Milk Vita milk pocket areas.

Livestock play an important role in human nutrition, directly through the consumption of animal products by livestock owners and their families and indirectly through the sale of animals and animal products to provide a source of income. The offtake of livestock and sale of surplus livestock products enable poor rural families, in particular women, to enter the cash economy. Small-scale producers do not use food that can be used for human consumption to feed livestock. The contribution of livestock in providing food security has seldom been examined, and the role of livestock in food security is highly undervalued (BBS 2016).

4 What is Hydroponics?

In natural conditions, soil acts as a mineral nutrient reservoir, but the soil itself is not essential to plant growth. When the mineral nutrients in the soil are dissolved in water, plant roots are able to absorb them. When the required mineral nutrients are introduced into a plant's water supply artificially, soil is no longer required for the plant to thrive. Almost any terrestrial plant can grow like this. This method of growing plants using mineral nutrient solutions, in water, without soil is known as hydroponics. It is possible by hydroponic techniques to achieve better than normal farm production, immune to natural weather variations, as well as organic and more nutritive, in just about 5% of the space and 5% of the irrigation water. NASA is reported to be working on this subject to meet fresh green food needs in space (Marckowiak et al. 1989).

Today, hydroponics is an established branch of agronomy. Progress has been rapid, and results obtained in various countries have proved it to be thoroughly practical and to have very definite advantages over conventional methods of horti-culture. There are two chief merits of the soilless cultivation of plants. First, hydroponics may potentially produce much higher crop yields. Also, hydroponics can be used in places wherein ground agriculture or gardening is not possible (New Zealand Merino Company 2011).

5   Need of Green Fodder for Cows

Green fodder is the natural diet of cattle. Green fodder is the most viable method to not only enhance milk production but to also bring a qualitative change in the milk produced by enhancing the content of unsaturated fat, omega-3 fatty acids, vitamins, minerals, and carotenoids.

Hydroponics fodder growing is the state-of-the-art technological intervention to supplement the available normal green fodder resources required by the dairy cattle. But, after the unfortunate Fometa experience, Indian scientists and planners have not given any attention to this subject. With increased pressure on farmlands to produce increasing needs of food grains, providing green fodder by hydroponics fodder growing is a necessity for the Indian dairy industry.

Modern researches have confirmed that grass-fed cow's milk is very rich in essential fatty acids (EFAs). Omega-3 is the most important constituent of grass-fed cow's milk, particularly for the brain and eyes. Some clinical studies indicate that a 1:1 ingested ratio of *omega*-6 to *omega*-3 (especially linoleic vs alpha-linolenic) fatty acids is important to maintaining cardiovascular health.

Typical Western diets provide ratios of 10:1–30:1 (i.e., dramatically higher levels of $n - 6$ than $n - 3$). The ratios of $n - 6$ to $n - 3$ fatty acids in common vegetable oils are canola 2:1, soybean 7:1, olive 1:33, sunflower (no $n - 3$), flax 1:3, cottonseed (almost no $n - 3$), peanut (no $n - 3$), grapeseed oil (almost no $n - 3$), and corn oil 46:1 ratio of $n - 6$ to $n - 3$. When a cow is raised on pastures, her milk has an ideal one to one ratio of these two EFAs discussed above. Studies suggest that if your diet contains roughly equal amounts of these two fats, you will have a lower risk of cancer, cardiovascular disease, autoimmune disorders, allergies, obesity, diabetes, dementia, and various other mental disorders.

6   Why Hydroponic Fodder?

In Bangladesh, the demand for green fodder is increasing on the account of the diversified use of agricultural residues. Adequate attention is not being given to production of fodder crops due to the increasing pressure on land for production of food grains, oil seeds, and pulses. In order to meet this increasing demand for green fodder, the next best alternative is hydroponic fodder to supplement the meager pasture resources. Some of the benefits of hydroponic fodder production are as follows:

- Land preservation
- Water conservation
- Faster growth and maturity
- Contamination-free
- Minimal use of fungicide and pesticide
- Less labor and maintenance costs

- Control over growing environment
- Time-saving
- Continual produce
- Weed-free
- Highly palatable and nutritious fodder

7 Introduction of Solar Power to Hydroponics

A simple hydroponic fodder machine which can produce up to 270 kg of product every day was taken as a case study for land and electricity feasibility study.

Generally, these machines are made of aluminum structures which are cemented in the ground. The overall dimension of the whole system structure is as described below:

$$\text{Systems structure dimension} : 18 \text{ ft (long)} \times 10 \text{ ft (width)} \times 8 \text{ ft (height)}$$
$$= 1440 \text{ cubic ft (volume)}$$
$$= 180 \text{ sq ft (area)}$$

This structure is cemented in the floor so it requires few more feet for installation.

$$\text{Actual area on the floor} = 20 \text{ ft (long)} \times 14 \text{ ft (width)}$$
$$= 280 \text{ sq ft}$$

The amount of product that it produces 270 kg is sufficient to feed 10–25 cows or up to 70 full-grown goats.

Now, if we take a look at the current scenario of the traditional features in terms of the grazing land and no. of animals, there are two main aspects that need to be taken into consideration:

(a) How many animals should be on your pasture?

Limited amount of land but a flexible herd size, you probably want to know the maximum number of animals that you can graze on your pasture.

(b) *How many acres of pasture do your animals need?*

If you have a lot of land but you want to keep a fixed number of livestock, you probably want to know the minimum amount of land that your animals need to graze.

To answer the questions above, you'll need to know:

- No. of days of your grazing season
- Weight of your animals
- Acres of land available for grazing
- Standard yield of your pasture per acre
- The rate of livestock used

For proper demonstration let's discuss the matter with an example. If we consider to figure out the number of animals and how much land is required for the whole year (365 days), the following data are taken as an arbitrary basis:

- The standard weight of one beef cow/calf is 1000 lb.
- Total of 20 acres of pasture.
- Together, the average yield of pastures is 10,000 lb./ac.
- The rate of livestock used. This number is generally 0.04 or 4%. This is used as a number from the thump method yielding to the fact that livestock have to take 4% of their weight each day (New Zealand Merino Company 2011).

Total number of animals

$$= \text{Total of 20 acres of pasture} \times \text{average yield of pasturesis } 10,000 \text{ lb/ac}(0.04)$$
$$\times (\text{average animal weight } 1000 \text{ lb}) \times (\text{grazing days } 365) = 13$$

Now let's figure out the minimum amount of pasture our animals would need. Let's use the 13 beef cows from above.

Acres of land required

$$= \frac{(\text{number of animals}) \times (\text{average animal weight}) \times (\text{rate of livestock used}) \times (\text{grazing days})}{\text{average yield per acre}}$$
$$= \frac{13 \times 1000 \text{ lb} \times 0.04 \times 365}{10,000 \text{ lb}}$$
$$= 9 \text{ acre}$$
$$= 836,000 \text{ sq feet}$$

As we can see from the calculation, we need about 836,000 sq. feet land to feed 13 cows annually, whereas about 280 sq. feet land is good enough to produce 270 kg of food everyday. But the obvious thing that was taken out of the equation is the electricity that will be required to produce that amount of food (Hydroponic fodder production 2011).

7.1 Electricity Supply

After calculating the land requirement, the electricity source, quality, and its cost for a fodder machine were analyzed. The specifications of a typical fodder machine are as described below.

The major electrical components of fodder machines are:

1. An air conditioner
2. A blower
3. A small water pump, etc.

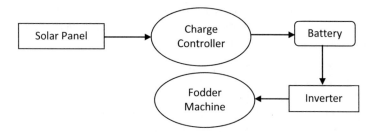

Fig. 1 Schematic diagram of solar-powered stand-alone system

In a typical fodder machine (18 ft. (long) × 10 ft. (width) × 8 ft. (height)), electricity connected load is 1.5–2 kW with continuous supply and 12,000–14,000 BTU. Six hundred liters of water is needed daily [8].

From our previous experience in disseminating the fodder machine technology in rural areas of Bangladesh, we have found that a major obstacle of the implementation was the quality of electrification. In Bangladesh, though 72% of the population have access to the electricity, the voltage fluctuation and continuous load shedding are major problems of the rural electrification. As a result, this has hindered new technologies such as hydroponic fodder machine to be applied by the farmers and businessmen.

In order to solve this problem, solar-powered hydroponic fodder machine is proposed in the present study. The system can be installed either as a stand-alone or a grid-connected system.

7.1.1 Solar-Powered Stand-Alone System

This system is efficient and suitable for rural areas, especially where grid electrification is not available. The system consists of a solar panel, a charge controller, a set of battery, and an inverter (Fig. 1).

7.1.2 Solar Grid-Connected System

A solar grid-connected system has advantages in terms of cost benefit and sustainability. The gridline connection will help the system life span and reduce per kg of food production costing.

As the gridline connection will ensure supply of electricity into the system, the system will not be in operation while power is supplied from the gridline. As a result the total running time of the machine will reduce on a given day. It also means that the maintenance or shutdown will also be delayed and battery will run for more months than the regular continuous use. So it will also be profitable if we consider the costing of the maintenance and new battery installation.

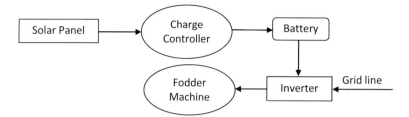

Fig. 2 Schematic diagram of solar grid-connected system

This system contains a solar panel, charge controller, battery, inverter, and connection from the grid line to the inverter (Fig. 2).

8 Summary and Conclusions

There are three important factors that can be summarized in this present work:

1. Land: In a populated country like Bangladesh where there is acute shortage of land and the population is around 160 million, fodder machine can certainly play a vital role. According to the World Bank report of 2014, the total cultivable land is 69.9% resulting into 90,990 sq. km. But these areas of land are used mainly for producing crops not as the grazing land. The abandoned farming land after harvesting the crops are mainly used as the pasture. The fodder machine can be a huge opportunity to fill it as only 280 sq. is good enough to feed the 10–12 full-grown cows.
2. Power: According to the report of the World Bank in 2014, 51.4% of people have electricity access. So almost half of the population doesn't have the access to the electricity. Obviously, access to the electricity doesn't ensure the quality. This solar-powered hydroponic fodder machine runs only 1.5–2 KW power. This on or off grid system can be a good solution to the electricity problem as the remaining power can also be used for home system. The economic viability also stands strong as the fodders have high demand in the market provided Bangladesh being an agro-based country. It also ensures the clean energy and works in favor of government who aims to achieve 10% of the total electricity from the source of renewable energy.
3. Nutrition: Cattles grazing from the abandoned pasture of harvested farming land don't get enough nutrition. But the products from the fodder machine can produce fodder which contains protein, vitamin B1, iron, potassium, magnesium, and zinc. This is able to provide the cattle with proper nutrition.

Thus, the solar-powered hydroponic fodder machine can be a solution to the nutrition of the cattle, a secondary power solution and an ultimate solution to the grazing land. In addition, this technology has an immense prospect in Bangladesh due to the fact that Bangladesh is an agro-based country. Furthermore, Bangladesh is also among the emerging countries who had attained the Millennium Development Goals (MDGs) and now is moving toward attaining the Sustainable Development Goals (SDGs) by 2030. This type of machine can help to attain SDG-1, 7, 8, and 9. This machine is a new income source and will encourage new entrepreneurs which will reduce the poverty, yield to provide new jobs, and help the economic growth of the country, aligning with goal nos. 1, 8, and 9. This is a form of clean energy as it is backed by solar power which aligns with goal no. 7. Hence, solar-powered hydroponic technology is in line with the government's policy direction and helps Bangladesh toward sustainable development.

References

BBS (2016) Yearbook of Agricultural Statistics-2015 27th Series Bangladesh Bureau of Statistic, Dhaka

Bhattacharjee B, Khatun H (2016) Brief on livestock. Statistics Department of Livestock Services, Dhaka. [access date: 01.03.18] http://www.fao.org/fileadmin/templates/rap/files/meetings/2016/160801_BGD-SurveyCalendar__Brief_of_Livestock_Statistics.pdf

FAO (2005) Brief on Livestock Statistics: Bangladesh. Food and Agriculture Organization of Bangladesh, Rome

Huque KS, Sarker NR (2014) Feeds and feeding of livestock in Bangladesh: performance, constraints and options forward; Bangladesh. J Anim Sci 43(1):1–10

Hydroponic Fodder Production (2011) An analysis of the practical and commercial opportunity, the New Zealand Merino Company. http://hulp.landbou.com/wp-content/uploads/2014/03/AIG_Grant_1122_Merino_NZ_Hydroponic_Fodder_Production.pdf

Marckowiak CL, Owens LP, Hinkle CR, Prince RO (1989) Continuous hydroponic wheat production using a recirculating system. National Aeronautics and Space Administration, Florida

New Zealand Merino Company (2011) Hydroponic fodder production: an analysis of the practical and commercial opportunity. New Zealand Merino Company, Christchurch

Salim Hossan MD (2017) Livestock Economy at a Glance 2015–16. Department of Livestock Services, Dhaka. [access date 01.03.18] http://dls.portal.gov.bd/sites/default/files/files/dls.portal.gov.bd/page/5f7daa39_d71f_4546_aeaf_55b72ee868f2/Updated%20Livestock%20Economy%20%282015-2016%29.pdf

An Approach to Optimize Cultivable Land Use for Solar PV Installation

S. M. Sanzad Lumen and Md. Zakirul Islam Sarker

1 Introduction

Development of renewable energy is one of the important strategies adopted as part of fuel diversification and energy security program. The International Energy Agency projected that solar power could provide a third of the global final energy demand after 2060, while CO_2 emissions would be reduced to very low levels (International Energy Agency Publication 2011). Solar photovoltaic (PV) has huge potential as a source of renewable energy, and it is going to play a vital role in power generation and climate change mitigation in the coming days.

With the current available technology, the power conversion efficiency of PV panel is still very low. That is why huge amount of land is required for large-scale PV power generation. But in a small country like Bangladesh, the scarcity of land is acute for such type of applications. PV plants require huge amount of land. Acquisition of such amount of land is difficult, and this also increases the initial investment of the project. Furthermore, once the land is used for PV plant installation, the land no longer can be used for cultivation. That is why cultivable land for conventional PV plant is discouraged. Moreover, in most of the cases, fixed tilted panel topology is used for the large-scale PV installation. Fixed tilted topology lacks optimal utilization of the land. To solve these problems, the whole PV system needs to be designed in such a way so that the land can be used for both cultivation and power generation. Also the arrangement of the panels needs to be furnished in an effective way so that maximum number of panels can be installed within a smaller area.

S. M. S. Lumen (✉)
Directorate of Renewable Energy and Research & Development, Bangladesh Power Development Board (BPDB), Dhaka, Bangladesh

Md. Z. I. Sarker
Directorate of Design & Inspection-1, Bangladesh Power Development Board (BPDB), Dhaka, Bangladesh

In Bangladesh, land is mainly used for cultivation. Some of them are used for cultivation throughout the year; some are half of the year and some are used even less than 6 months. In the remaining period of the year, there may be no use of it or minimum use for vegetables or other minor cultivation. This is because of the land type, local climate, insufficient rain, and low water table, etc. Also, some lands are not fertile enough to grow crops twice or thrice a year. These types of lands are cultivated mainly during rainy season due to irrigation water availability. However, the total area under cultivation is almost 22.98 million acres, which is almost 63% of the total land area. Among them, 9.14 million acres of lands are under irrigation. While 1.11 million acres of lands are single cropped and are cultivated half a year or less in most of the cases (Rahman 2008).

Bangladesh is located in the tropical monsoon region, and its climate is characterized by high temperature, heavy rainfall, often excessive humidity and fairly marked seasonal variations. From the climatic point of view, three distinct seasons can be recognized in Bangladesh – the cool dry season from November through February, the pre-monsoon hot season from March through May and the rainy monsoon season which lasts from June to September (National Encyclopedia of Bangladesh). Considering this climatic condition and type of land, a flexible tracking-based solar PV system design has been proposed in this paper. Lands which are used for 6 months or less for cultivation have been considered for this study. During the rainy season, the land will be used for cultivation, and the solar panels will be rotated east to west in such a way so that the crops are not shaded. While during the remaining period of the year, the panels will be on sun-tracking mode to receive maximum solar radiation. Thus, the land will be used for both power generation and cultivation.

2   Methodology

The methodology incorporates new ideas for dual usage of land for cultivation as well as for power generation. PV panels will be mounted on a structure which will allow the panels to rotate themselves vertically from east to west using one axis tracking system. Thus, during the power generation period, the panel will be faced to the sun and will be automatically rotated to track the sun. In this case total solar radiation will be used to generate power. Moreover, due to the deployment of one axis tracker, the overall energy output of the system will be increased by 13% than that of the tracker less system (Mageshkannan et al. 2013).

The PV plant will be designed to maximize the power production during months from October to May. The panels' tilt angle is kept zero, and hence no shading will be imposed on the adjacent panels due to seasonal variation of sun position, and also the panels will track the sun from sunrise to sunset to collect maximum solar radiation.

Besides, during the cultivation period the panel surface will be maintained to be parallel with the incident sun ray in order to keep minimum shading on the crops.

This time the panel edge will track the sun and hence keep minimum shading on crops. During this period some part of the diffused solar radiation will be used to generate power.

The system will be a grid-connected system, and all the generated electricity will be exported to the grid at FIT (feed-in tariff) or any contracted price. A contract will be signed between the project owner and the farmer/landowner for PV panel installation on that land. The landowner/farmer will be offered financial packages of certain percentage of revenue earned from electricity sale to grid. The more the electricity generation, the more the landowner/farmer will be financially benefitted. Besides, farmers can cultivate the land and use the generated electricity for irrigation whenever needed, while the project owner will get the land for almost free of cost and will set up a large PV power generation plant.

3   Design and Drawing

In this study, 4 acres of single cropped land mainly cultivated in between June and September is used to evaluate the proposed design as well as for the financial analysis. The analyses and evaluations were carried out using the RETScreen software. The proposed project area which consists of four pieces of land is shown in Fig. 1. Each piece of land has an area of 4050 m² or approximately 1 acre. In the present analysis, a solar module with capacity of 300 W_p has been considered, and the specifications are as listed in Table 1.

The panel structure is designed in such a way so that each of the structure can hold 24 pcs of PV modules. Structure to structure spacing will be 0.5 m, and several structures will be mounted on 1 acre of land. The details of the structure are described in Table 2.

The proposed design is shown in Fig. 2. The modules are mounted horizontally (tilt angle is 0°), and they can be rotated automatically along east to west using solar tracking system. The mounting structure is 8.5 meter long, 7 meter wide and 4 meter

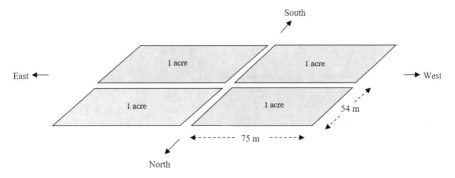

Fig. 1 Land areas in this study

Table 1 PV module rating

Sl. No.	Description	Rating
1	Type of solar cells	Polycrystalline
2	Rated power/maximum power (P_{max})	300 W
3	Open-circuit voltage (V_{oc})	43.8 V
4	Short-circuit current (I_{sc})	9.10 A
5	Voltage at P_{max} (V_{mp})	36.25 V
6	Current at P_{max} (I_{mp})	8.28 A
7	Power conversion efficiency of PV module	15%
8	Front cover	High transmission tempered glass
9	Static load front and back (e.g. wind)	2400 Pascal
10	Front loading	5400 Pascal
11	Weight and dimension	24.5 kg, 1958 mm*992 mm*50 mm

Table 2 Structure details

Sl.	Structure dimension	Structure leg	Module holder
1	Length: 8.5 m Width: 7 m Height: 4 m	MS pipe Diameter: 3″	MS pipe Diameter: 2″

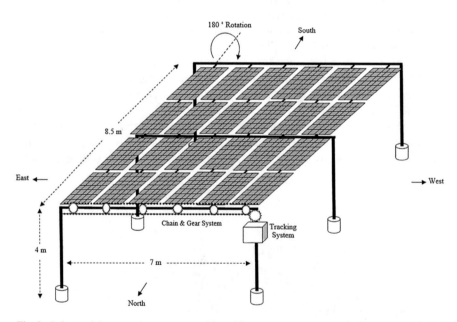

Fig. 2 Solar module, mounting structure, and tracking system arrangement in the proposed design

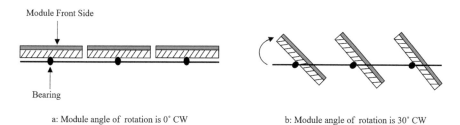

a: Module angle of rotation is 0° CW   b: Module angle of rotation is 30° CW

Fig. 3 Cross-sectional view of the solar module rotating mechanism viewed from north

high above the ground. The modules on the same axis (along the length) will be coupled to the main gear of the tracker through chain pulley system.

Module rotating mechanism is shown in Fig. 3. The sun tracker will track the position of the sun in 30 min interval and align the module accordingly. As the tracking system will track the sun in 30 min interval instead of frequent tracking, this will keep the tracker consume at a low level of energy.

4 Operation Profile

The system will be operated based on two profiles. During the cool dry season from November through February and the pre-monsoon hot season from March to May, it will be operated according to noncultivation profile, while during the rainy monsoon season which lasts from June to September, it will be operated according to cultivation profile. The two profiles are described below.

4.1 Noncultivation Profile

During noncultivation period, power generation will get the maximum priority. So, the solar panel alignment will be maintained in such a way so that the panel front side always gets the maximum solar radiation. In this profile the solar panel front side will always track the sun so that the incident sun ray is perpendicular to the panel front side. The duration of the noncultivation profile will be from October to May. During this period the direct solar radiation will be the major contributor to the generated power, and also the plant energy yield will be maximum. The cyclic operation of the system in a day during this profile is as shown in Fig. 4.

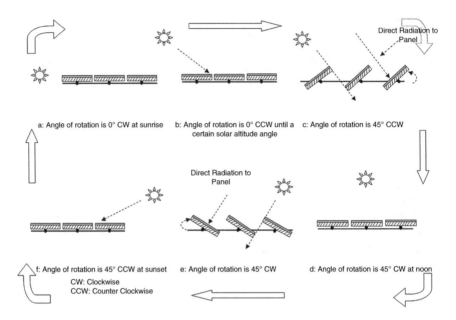

a: Angle of rotation is 0° CW at sunrise   b: Angle of rotation is 0° CCW until a   c: Angle of rotation is 45° CCW
                                            certain solar altitude angle

f: Angle of rotation is 45° CCW at sunset   e: Angle of rotation is 45° CW   d: Angle of rotation is 45° CW at noon
   CW: Clockwise
   CCW: Counter Clockwise

Fig. 4 Module orientation in a day during noncultivation profile

4.2 Cultivation Profile

During the cultivation period, crop cultivation will get the maximum priority, and power generation will be the secondary objective. So, the solar panel alignment will be maintained in such a way so that the crops get the direct solar radiation, and minimum shading is imposed on the crops.

In this profile the solar panel front side will be always kept to be parallel with the incident sun ray. In other words, the panel edge will track the sun and keep the angle between panel front side and incident ray at 0°. As a result, the crop cultivation will not be hampered though solar panels are installed on that land. Part of the diffused solar radiation will be used to generate electricity during this period, and hence, the PV plant will operate at a low capacity condition. Nonetheless, during monsoon season, where the sky is cloudy, the diffuse radiation is also higher in this period, and this favours the power generation. The cyclic operation of the system in a day during cultivation profile is as shown in Fig. 5.

When the sky is clear and the sun is very high in the sky, direct radiation is around 85% of the total insolation striking the ground, and diffuse radiation is about 15%. As the sun goes lower in the sky, the percent of diffuse radiation keeps going up until it reaches 40% when the sun is 10° above the horizon. Atmospheric conditions like clouds and pollution also increase the percentage of diffuse radiation (Exploring Science and Technology) (Fig. 6).

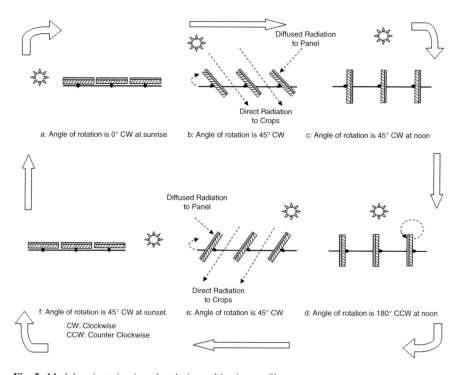

Fig. 5 Module orientation in a day during cultivation profile

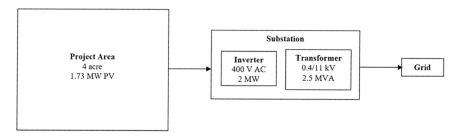

Fig. 6 System block diagram

5 Calculation

The module arrangement and structure are shown in Fig. 7. Each of the structures holds 24 PV modules. Taking into account the length and width of each of the modules and moderate module to module spacing, the dimension of the structure is selected as 8.5 m × 7 m, while structure to structure spacing is kept 0.5 m. So, each of the structure occupies 9 m × 7.5 m area.

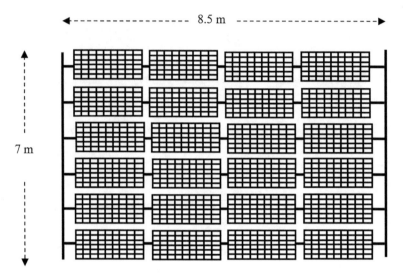

Fig. 7 Top view of a single structure

Area covered by a single structure, A = 9 m × 7.5 m = 67.5 m², while land area = 1 acre ≈ 4050 m².

Number of structure that can be mounted on 1 acre of land surface = 4050/67.5 = 60 nos.

Installable Panel Capacity on 1 Acre of Land

The configuration of panels at the field is shown in Fig. 8, and they are as follows:

No. of structure along length of the land = 75/7.5 = 10 nos
No. of structure along width of the land = 54/9 = 6 nos
No. of structure mountable on the land = 10 × 6 = 60 no
No. of solar module installable = 60 × 24 = 1440 pcs

∵ Installable capacity = 1440 × 300 Wp = 432,000 Wp = 432 kWp

Total Installable Capacity

Installable capacity on 1 acre of land = 432 kWp.
Total project area = 4 acre.

∵ Total Installable capacity = 432 × 4 = 1728 kWp = 1.728 MWp ≈ 1.73 MWp.

Electricity Export to Grid

Using one axis tracker, 0° tilt angle, 180° azimuth angle and 1.73 MW$_p$ PV panel capacity, the proposed design has been analysed in RETScreen. Table 3 shows the data of the solar radiation and electricity exported to grid.

The individual monthly electricity generation is also higher during noncultivation profile (Table 4). As the direct radiation is around 85% and diffuse radiation is about

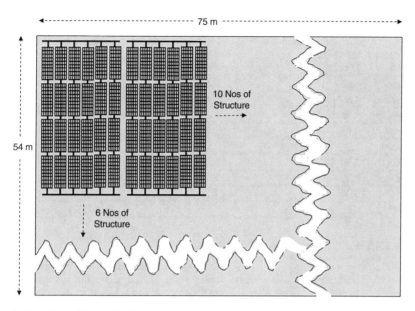

Fig. 8 Top view of the project land

Table 3 Daily solar radiation and electricity export

Month	Daily solar radiation – horizontal kWh/m²/day	Daily solar radiation – tilted kWh/m²/day	Electricity exported to grid MWh
January	4.35	6.06	246.1
February	5.22	7.55	271.1
March	6.10	8.31	322.2
April	6.20	7.80	291.4
May	5.74	7.12	276.0
June	4.77	5.51	208.1
July	4.19	4.69	184.1
August	4.29	4.83	189.3
September	3.89	4.48	171.2
October	4.67	6.25	246.3
November	4.66	6.46	249.4
December	4.26	6.09	246.6
Annual	**4.86**	**6.25**	**2901.9**

15% of the global insolation, considering other aspects, it is assumed that the diffuse radiation would be 10% of the global radiation, and the power generation by diffuse radiation would also be 10% of the total electricity generation. These assumptions are as shown in Table 5.

Table 4 Electricity export during noncultivation profile

Month	Daily solar radiation – horizontal kWh/m^2/day	Daily solar radiation – tilted kWh/m^2/day	Electricity exported to grid MWh
October	4.67	6.25	246.3
November	4.66	6.46	249.4
December	4.26	6.09	246.6
January	4.35	6.06	246.1
February	5.22	7.55	271.1
March	6.10	8.31	322.2
April	6.20	7.80	291.4
May	5.74	7.12	276.0
Total			**2149.1**

Table 5 Electricity export during cultivation profile

Month	Daily solar radiation – horizontal kWh/m^2/day	Daily solar radiation – tilted kWh/m^2/day	Diffused solar radiation – tilted kWh/m^2/day	Electricity exported to grid due to global radiation MWh	Electricity exported to grid due to diffuse radiation MWh
June	4.77	5.51	0.55	208.1	20.81
July	4.19	4.69	0.47	184.1	18.41
August	4.29	4.83	0.48	189.3	18.93
September	3.89	4.48	0.44	171.2	17.12
Total					**75.27**

Table 6 Project costing

Total initial costs	100.0%	$	.	6,138,881
Annual costs and debt payments				
O&M		$		20,000
Fuel cost – Proposed case		$		0
Debt payments – 25 yrs		$		421,424
Total annual costs		$		441,424

∵ Annual electricity export to grid = (electricity exported to grid during noncultivation profile) + (electricity exported to grid during cultivation profile) = 2149.1 + 75.27 = 2224.37 MWh ≈ 2225 MWh.

Financial Analysis

The financial analysis was carried out by using the RETScreen. The cost of the project, annual income and the financial viability are as shown in Tables 6, 7 and 8, respectively, while Fig. 9 shows the cumulative cash flow of the project.

Table 7 Annual income

Electricity export income		
Electricity exported to grid	MWh	2225
Electricity export rate	$/MWh	250.00
Electricity export income	$	556,179
Electricity export escalation rate	%	2.0%

Table 8 Financial viability

Pre-tax IRR – equity	%	23.6%
Pre-tax IRR – assets	%	1.3%
After-tax IRR – equity	%	23.6%
After-tax IRR – assets	%	1.3%
Simple payback	Year	10.5
Equity payback	Year	5.0
Net present value (NPV)	$	891,782
Annual life cycle savings	$/year	113,702
Benefit-cost (B-C) ratio		1.73
Debt service coverage		1.30
Energy production cost	$/MWh	206.5
GHG reduction cost	$/tCO2	(84)

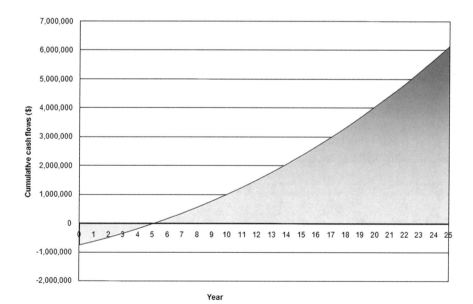

Fig. 9 Cumulative cash flow

Table 9 Comparison

Proposed design	Conventional design
Installed capacity: 1.73 MW	Installed capacity: 1.62 MW
Electricity exported to grid: 2225 MWh	Electricity exported to grid: 1775 MWh
IRR: 23.6%	IRR: 10.1%
Simple payback: 10.5 years	Simple payback: 13.3 years
Equity payback: 5.0 years	Equity payback: 12.1 years
Energy production cost: 206.5 $/MWh	Energy production cost: 258.84 $/MWh
Cultivation: Possible for 4 months	Cultivation: Not possible

6   Results and Discussion

In Bangladesh, during the summer period, the electricity demand increases due to cooling load and irrigation load. This increase in load puts huge pressure on the grid, and utilities have to frequently load shed to manage the generation shortage. The study shows that the PV-generated power will be maximum during the summer period; hence, it will boost the generation and will eventually mitigate the generation shortage, while the cultivation period basically falls in the rainy season when cooling load and irrigation load are less. Moreover, during the rainy season, rainwater will be useful for crop cultivation, and the higher diffuse radiation due to cloudy sky will be used for electricity generation in the cultivation period.

If the same amount of land was used for conventional fix tilted PV plant installation, the installable capacity would be 1.62 MW_p (10 m^2 area for 1 kW_p panel). A comparison between the proposed approach and the conventional approach has been carried out by using RETScreen, and the results are as shown in Table 9.

The proposed design ensures higher IRR and higher energy yield and thus offers less payback period. It also requires less investment as the land is not purchased. So, in all aspects, the proposed design is more viable than the conventional approach. Moreover, this design is technically and financially viable for Bangladesh in all aspects. It will bring positive impacts for the socio-economic development and will help in reducing the greenhouse gas emissions.

7   Conclusion

Simulations from this study show the potential of maximizing the power generation by installing PV plants on the less cultivated lands. The operational design in this study facilitates the dual use of land for power generation and for cultivation. In this way, the proposed design will ensure the optimum use of less arable land for the whole year. Not only would this help in overcoming the power generation shortage but also in reducing the carbon emission. This is in line with the Bangladesh's renewable energy target of generating 10% of total electricity production from renewable energy sources by 2020.

Future Scopes

In this paper the proposed design has been explained and analysed. But there is further scope to demonstrate and evaluate the design practically by a pilot project. Few issues like energy generation by diffuse radiation, design consideration of the solar tracker, suitability of two axis trackers, land plowing issues, contract norms between project owner and landowner, etc. are to be evaluated and resolved before going for commercial-scale projects.

References

Mageshkannan P, Deepthi S, Ponni A, Ranjitha R, Dhanabal R (2013) Comparison of Efficiencies of Solar Tracker systems with static panel Single-Axis Tracking System and Dual-Axis Tracking System with Fixed Mount. International Journal of Engineering and Technology (IJET) 5(2) 1925–1933

Exploring Science and Technology. Accessed on 23 June 2017. Website: http://www.ftexploring.com/solar-energy/direct-and-diffuse-radiation.htm

Rahman Md. H (2008) Agricultural land use and land susceptibility in Bangladesh: an overview introduction. Website: https://www.researchgate.net/profile/Hasibur_Rahman4

International Energy Agency Publication. (2011) Renewable Energy Technologies: Solar Energy Perspectives, pp 19–22. Website: http://www.iea.org/Textbase/npsum/solar2011SUM.pdf

National Encyclopedia of Bangladesh. Accessed on 15 June 2017. Website: http://en.banglapedia.org/index.php?title=Climate

Large-Scale Solar-Assisted Water Heater for a Green Hospital

Poorya Ooshaksaraei, Khalid Mokhtar, Syed Zulkifli Syed Zakaria, and Kamaruzzaman Sopian

1 Introduction

Water heating is one of the most well-known and common applications of solar energy because of its feasibility and economic advantages compared with other solar energy applications. Solar water heating is cost-efficient in domestic and industrial applications (Islam et al. 2013). From 2005 to 2010, the installed capacity of solar water heaters (SWHs) has increased by 16% per year (Islam et al. 2013). SWH systems, which are preferably operated at a temperature range of 55–95 °C for industrial, commercial, and domestic applications, have various advantages (Gang et al. 2012). Numerous studies have been conducted on various SWH designs to evaluate and improve their performance in residential and industrial applications (Mekhilef et al. 2011). SWH units are used in residential buildings in the equatorial region of Malaysia. Other industrial and commercial applications require massive amounts of low-temperature thermal energy, thus implying the promising use of large-scale SWH systems in these sectors (Mekhilef et al. 2012).

A solar-assisted water heating system was proposed for use at Hospital Universiti Kebangsaan Malaysia (HUKM) in 2010. This hospital aims to develop into one of

P. Ooshaksaraei (✉)
Solar Energy Research Institute, Universiti Kebangsaan Malaysia (UKM), Bangi, Selangor, Malaysia

Mehr Pars Renewable energy Development Co., Rasht, Guilan, Iran

K. Mokhtar
Zamatel Sdn. Bhd., Wisma MAIS, Shah Alam, Selangor, Malaysia

S. Z. Syed Zakaria
UKM Medical Center, Wilayah Persekutuan Kuala Lumpur, Malaysia

K. Sopian
Solar Energy Research Institute, Universiti Kebangsaan Malaysia (UKM), Bangi, Selangor, Malaysia

© Springer International Publishing AG, part of Springer Nature 2018
H.-Y. Chan, K. Sopian (eds.), *Renewable Energy in Developing Countries*,
Green Energy and Technology, https://doi.org/10.1007/978-3-319-89809-4_8

the most modern, state-of-the-art, and green hospitals in Malaysia. The installation of the first large-scale SWH in Malaysia aims to achieve the following:

- Increase public awareness regarding the benefits of solar thermal energy
- Demonstrate the use of solar thermal energy in water heating as an economical alternative to fossil fuels
- Provide an advanced and practical platform for solar thermal research
- Evaluate the performance of solar thermal collectors as water preheaters in existing commercial hot-water services
- Reduce the amount of fossil fuels used and the emission of greenhouse gases (as presented in a case study)

The entire solar water heating system is installed at the ground level rather than on the rooftop to raise public awareness, which was a project objective.

This paper presents the SWH system design simulated using TRNSYS software (Klein 1979), a comparison of theoretical and experimental operation performances, and the economic and environmental benefits of the large-scale SWH.

2 Solar Irradiation at the HUKM Installation Site (A Case Study)

The monthly average daily solar irradiation obtained from 4 to 8 h of sunshine in Malaysia ranges from 4.7 to 5.4 kWhr/m^2 (Fig. 1) (Sopian and Othman 1992). The Malaysian sky is neither clear nor overcast (Zain-Ahmed et al. 2002), and Malaysia receives a considerable amount of rainfall. This climate significantly decreases irradiation (Sopian and Othman 1992) but washes away dust and cleans the solar panels. Frequent rainfall ensures the maximum absorption of available solar

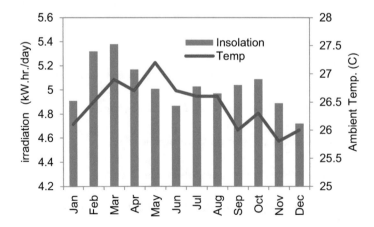

Fig. 1 Monthly average ambient temperature and irradiation in Malaysia

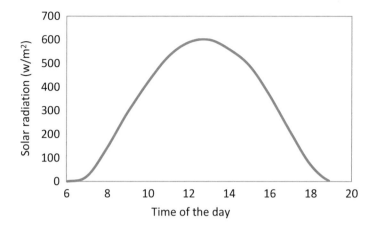

Fig. 2 Annual average daily solar radiation in Malaysia

Fig. 3 Satellite views of
HUKM and the allocated
SWH installation site
(googlemap)

radiation by the solar panels. Figure 2 shows the annual average daily solar radiation
intensity during a typical day in Malaysia.

Large-scale solar water heating applications have enormous potential in residential, commercial, and industrial sectors. Solar-assisted water heating systems can be
implemented in over 100 hospitals and hotels in Malaysia.

HUKM is a modern, state-of-the-art teaching hospital at the Faculty of Medicine
in UKM. HUKM is located in Cheras (Fig. 3), which is approximately 10 km from
the Kuala Lumpur City Centre at altitude of 3.03 and a longitude of 101.26
(itouchmap.com). HUKM is equipped with 1000 beds.

Solar panels are preferably installed on building rooftops to avoid shading from
surrounding buildings. However, an open area measuring 850 m^2 at the ground level
of HUKM was allocated for panel installation of the SWH because public awareness
was an important objective. The solar panel field is open to public visitors. Figures 3
and 4 show the site prior to installation.

Fig. 4 Allocated site prior to installation (ground level)

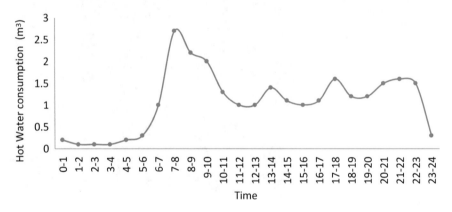

Fig. 5 Daily hot-water consumption at HUKM

3 Description of Hot-Water Demand at HUKM

Hot-water consumption at HUKM hospital has been closely monitored since 2010. A flow meter monitors LPG (liquefied petroleum gas) supply line to gas burners, and another flow meter monitors hot-water supply to evaluate the water consumption pattern. The hot-water consumption pattern is a key parameter in solar water heating systems design. The hospital consumes 26 m^3 hot water per day according to the meter reading. Figure 5 shows daily hot-water consumption at HUKM.

About half of the 26 m^3 daily hot water is required from 10 A.M. to 6 P.M., while there is a considerable solar radiation. So, the SWH meets total hot-water demand in this period. There is a spike in hot-water consumption in early morning where there

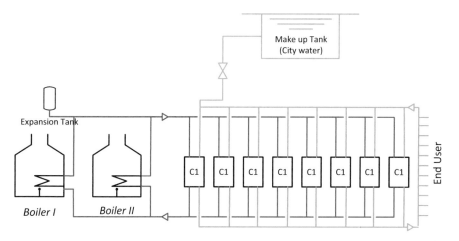

Fig. 6 The existing LPG burners' water heating system

is no sufficient solar radiation to heat up water. In this regard a properly sized hot-water storage tank is required to store, maintain, and deliver hot water at the required temperature.

The former water heating system generates hot water by two LPG burners and heat calorifiers. Each boiler has a capacity of 2.1 million kcal/h, and each calorifier has a capacity of 13,500 l/h. The system was composed of two loops (Fig. 6):

Primary loop (red color): Water circulates within the boiler and the calorifier in a closed loop. The expansion tank and water replenishment system compensate for the water lost during this closed-loop circulation.

Secondary loop (green color): This loop is composed of calorifiers and end-user distribution pipelines. A makeup tank replaces the water consumed by the end user during circulation.

Each calorifier is assigned to a network of users such as laundry, kitchen, wards, and toilets. The water consumed during circulation is then replenished by the water stored in the make-up tank. All eight calorifiers operate simultaneously based on demand from the end users.

The temperature of the water distributed to the user must range from 50 to 60 °C. The temperature of the hot water fed into the calorifier from the boiler, which is the primary hot-water supply, must be 90 °C. The makeup tank contains city water. The calorifiers end a signal to call in hot water at 90 °C when it detects a temperature drop below 50 °C. The boiler is simultaneously triggered until water flow in the secondary hot-water supply loop reaches the preset temperature of 60 °C. One boiler operates regularly and typically remains in hot-standby mode when it is not in operation. The second boiler typically remains in cold-standby mode and acts as a backup in case the first boiler fails.

4   U-Pipe Type Evacuated Solar Panel

Flat-plate solar thermal panels are widely used in water heaters in single-family residential buildings in Malaysia. However, evacuated panels are rarely installed. Evacuated panels are feasible for large-scale applications, especially industrial applications with high-temperature requirements, because they retain heat (Ma et al. 2010).Thus, evacuated tube panels show better thermal performance compared with flat-plate panels at high temperatures (Duffie and Beckman 1980; Foster et al. 2009).

The large-scale SWH project at HUKM aims to visualize and emphasize the benefits of large-scale solar thermal water heating. An evacuated solar thermal panel was selected to facilitate the applicability of the system in the industrial and commercial sectors, such as hospitals, hotels, and food industry.

U-pipe and heat pipe types are the most well-known evacuated tubes. While heat pipe solar panels operate at optimum when installed at a 20–40° slope (Elmosbahi et al. 2012), they are more applicable in high latitude rather than equatorial areas. Malaysia is close to the equator. Therefore, the optimum panel slope at HUKM is about 3°. But it is tilted 15° due to self-cleaning purpose. Several theoretical and experimental studies have been conducted on the thermal performance of solar U-pipe type evacuated tubes (Ma et al. 2010; Liang et al. 2011, 2013; Kim and Seo 2007).

A U-type pipe evacuated panel was selected from among a wide range of evacuated tube panels. Each panel contains 16 glass tubes. The U-pipes are made of copper and are welded to a copper manifold. The glass tubes are made of borosilicate glass, which is clear and transmits over 90% of solar radiation. The glass tube is evacuated to 1 MPa to eliminate thermal convection heat loss through the glass. The absorber is magnetron-sputtered with an aluminum nitride selective coating, which converts over 92% of incoming solar radiation into heat. The thermal emission coefficient is less than 8%. Table 1 lists the geometric characteristics of the selected collector.

Table 1 Geometric characteristics of the collector

Dimension(mm)	$1978 \times 1636 \times 134$
Weight (kg)	61.6
Absorber area (m^2)	1.297
Glass tube	
Outer diameter (mm)	Ø58 ± 0.7
Inner diameter (mm)	Ø47 ± 0.7
Length (mm)	1800
Heat loss ($w/m^{2\circ}C$)	≤0.6
U-pipe	
Outer diameter (mm)	Ø8
Inner diameter (mm)	Ø6.8
Header pipe	
Outer diameter (mm)	Ø15
Inner diameter(mm)	Ø13

5 Evaluation of Overall SWH Performance

Domestic and large-scale low-temperature applications of SWHs have been widely studied (Fisch et al. 1998). However, early research often evaluated the systems using energy balance methods based on the first law of thermodynamics. The first law of thermodynamics evaluates the system based on the quantities of input and output energy. The system must be evaluated using the exergy method, which is based on the second law of thermodynamics. The exergy method evaluates the quality of output energy (Xiaowu and Ben 2005).

5.1 First Law Efficiency

The first law of thermodynamics evaluates the ratio of output and input energies in a system. SWH releases thermal energy, which is delivered via hot water. Solar radiation on the collector surface provides input energy, and electrical energy is consumed by the circulation pumps as shown in Fig. 7.

SWH daily output thermal efficiency E_{Th} is defined as the transfer of thermal energy to working fluid:

$$E_{\mathrm{Th}} = \sum_{1}^{24} \dot{m}_i C_p (T_{\mathrm{out}} - T_{\mathrm{in}}) \tag{1}$$

where $\dot{m}_i$ denotes the total hot water delivered hourly to the hospital, C_p denotes the specific water heat, and T_{in} denotes the temperature of the water delivered to the hospitals and the temperature of the cold water that replenishes the SWH from the city water storage tank.

Daily input thermal efficiency of SWH is defined as the amount of solar radiation that reaches the solar panels throughout the day.

Fig. 7 Input and output energies of an active SWH

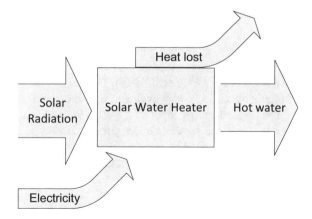

$$E_S = \sum_1^{24} S_i \times n \times A \tag{2}$$

where S_i denotes the hourly solar radiation, n denotes the number of solar panels, and A denotes the collector area of a single panel.

Electrical energy has a higher value than thermal energy, and this advantage must be considered in calculating total efficiency (Shahsavar and Ameri 2010). Daily electrical energy consumption of SWH is defined as the total electrical energy consumed by circulation pumps over the efficiency of conventional power plants.

$$E_{El} = \frac{\sum_1^{24} E_{El_i}}{\eta_{PowerPlant}} \tag{3}$$

where E_{El_i} is the electrical power consumed by circulation pumps per hour and $\eta_{PowerPlant} = 0.38$ is the conventional power plant efficiency (Ji et al. 2007).

Average daily energy efficiency of SWH can be defined as the ratio of SWH output energy to input energy.

$$\eta_E = {}^{E_{Th}}\!/_{(E_S + E_{El})} \tag{4}$$

Average hourly energy efficiency of SWH can be defined as the ratio of SWH output energy to input energy.

$$\eta_{Ei} = {}^{E_{Th_i}}\!/_{\left(E_{S_i} + E_{El_i}\right)} \tag{5}$$

where E_{Th_i}, E_{S_i}, and E_{El_i} denote average thermal energy output, solar radiation input, and electrical energy input per hour, respectively.

5.2 Second Law Efficiency

The first law of thermodynamics evaluates the quantity of thermal energy output; however, this measurement cannot perfectly evaluate the performance of the SWH system (Gang et al. 2012). The exergy method is based on the second law of thermodynamics, which evaluates both the quality and quantity of output energy.

The thermal energy input of a SWH (solar energy) is greater than the electrical energy consumed by circulation pumps. However, based on the second law of thermodynamics, thermal energy value is lower than the value of electrical energy.

Input exergy of SWH includes electrical and solar exergy (Fujisawa and Tani 1997):

$$Ex_{in} = Ex_{El} + Ex_S \tag{6}$$

Solar radiation exergy is given as (Petela 2003):

$$Ex_S = \sum_1^{24} S_i \left[1 - \frac{4}{3} \left(\frac{T_{a_i}}{T_{sun_i}} \right) + \frac{1}{3} \left(\frac{T_{a_i}}{T_{sun_i}} \right)^4 \right] \tag{7}$$

where T_{a_i} denotes the average hourly ambient temperature and S_i denotes the average hourly solar radiation.

Electrical energy can be converted into work efficiently, and thus electrical exergy is assumed to be equal to the electrical energy consumed by circulation pumps (Chow et al. 2009).

$$Ex_{El} = E_{El} \tag{8}$$

Thermal output exergy of SWH obtained in the storage tank (Chow et al. 2009) may be written as:

$$Ex_{Th} = \sum_1^{24} \dot{m}_i C_p \left[(T_{2_i} - T_{1_i}) \times \left(1 - {}^{T_{a_i}}/_{T_{2_i}} \right) \right] \tag{9}$$

where T_{a_i} denotes the ambient temperature, T_{1_i} denotes the city water temperature, T_{2_i} denotes the temperature of the hot water delivered to the hospital, and $\dot{m}_i$ denotes the hot water delivered to the hospital every hour.

Exergy efficiency is defined as output over the input exergies (Sarhaddi et al. 2010; Borel and Favrat 2010; Hepbasli 2008; Nayak and Tiwari 2008).

$$\eta_{Ex} = \frac{Ex_{Th}}{Ex_S + Ex_{El}} \tag{10}$$

Average hourly exergy efficiency of SWH can be defined as the ratio of SWH output exergy to input exergy.

$$\eta_{Exi} = \frac{Ex_{Th_i}}{Ex_{S_i} + Ex_{El_i}} \tag{11}$$

6   SWH Design

A SWH system is basically comprised of the following: solar collectors, storage tank, pumps, and a control system. The U-type pipe evacuated tube solar thermal collector directly uses the sun's energy to heat water. This collector typically consists of a U-pipe housed in a high-vacuum glass tube.

The final system design (Fig. 8) contains three loops, which are described as follows:

Energy harvest loop: This closed loop consists of the solar thermal collectors, a circulation pump, an expansion tank, and a water replenishment system. The heat energy is passed to the next loop via a heat exchanger.

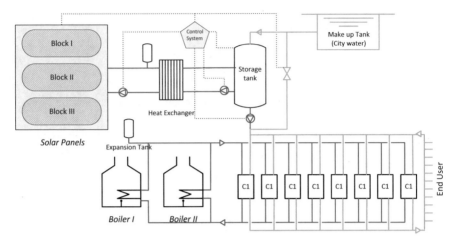

Fig. 8 Solar-assisted water heater design

Energy transfer loop: This open loop continues the transfer of heat from the heat
exchanger. It consists of a circulation pump and storage tank. The storage tank
damps the water expansions in this loop and replenishes water loss.

Water supply loop: This open loop replaces the water consumed by the end user via
the hot-water circulation loop.

The installation site for the solar collectors permits the installation of 144 panels;
however, this number is insufficient to fully support the hot-water requirements of
the hospital. Figure 8 shows a hybrid solar–LPG-assisted hot-water system. Water
supplied by the main water storage tank of the hospital passes to the SWH storage
tank. Subsequently, LPG boilers increase the temperature of preheated water to the
preset temperature.

The solar collector field is separated into three blocks connected in parallel. Each
block consists of 12 rows of panels, with each having 4 panels. The solar collector
module consists of 16 evacuated tubes. Thus, the SWH system is composed of a total
of 2304 evacuated tubes. The 4 panels in each row are connected in series, whereas
the 12 rows in each block are connected in parallel.

Splitting the system into 3 by 12 rows results in high tracing and evaluation
flexibility with respect to the individual performance of each block. These blocks are
connected in parallel to a heat exchanger.

Variable speed pumps were selected for this pilot project because this project is
intended to provide a research platform for researchers at the solar energy research
institute of UKM. A plate heat exchanger was also selected to increase the number of
plate and heat transfer surfaces. Water flow rate from the variable speed pumps can
be adjusted to evaluate system performance. A certain range of total flow rate of the
"energy harvest loop" can be adjusted by manipulating the inverter of pump 2. Each
loop is equipped with a valve that reduces flow rate for research purposes.

7 Simulation Using TRNSYS Software

A forced circulation solar water heating system is simulated using TRNSYS software. TRNSYS is an extensive, well-known quasi-steady-state simulation software developed at the University of Wisconsin (Xiaowu and Ben 2005). It is widely adopted both commercially and academically (Carrillo Andrés and Cejudo López 2002; Calise et al. 2013; Ayompe et al. 2011). System performance is dynamically examined. Figure 9 presents the visualized configuration of TRNSYS software components.

The system configuration described above is modeled on the available climate data of Malaysia. Annual storage tank temperature is determined using TRNSYS software and is shown in Fig. 10. The temperature of the hot-water storage tank is expected to range from 35 to 95 °C based on simulation results. This temperature range is acceptable for the majority of hot-water end users at the hospital. A downsized or oversized storage tank causes the system to either overheat or provide low-temperature hot water.

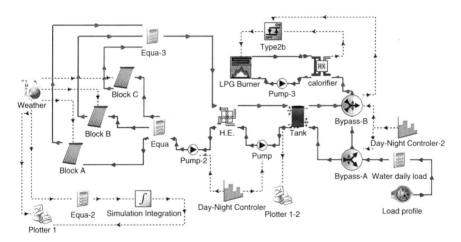

Fig. 9 TRNSYS simulation of the final SWH system design

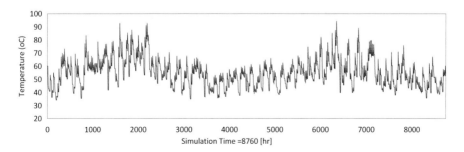

Fig. 10 Annual temperature in the hot-water storage tank

Energy loss and water temperature drop along the piping system are neglected. A 15-ton, nonpressurized storage tank stores and supplies hot water to end users. Hot-water consumption in the hospital is low during the weekend, and the storage tank is designed to save supplementary energy for the following day. The storage tank is made of stainless steel with 50 mm stone wool-type insulation, and its capacity must be downsized to minimize the amount of hot water stored in the tank while preventing system overheats. Daily temperature variation in the storage tank is determined based on Fig. 8. The annual average is listed beside the highest and the lowest monthly average daily temperatures in the storage tank.

Hybrid water heating operation is divided into two modes:

- 10 A.M. to 10 P.M.: Hot water is generated by the hybrid SWH.
- 10 P.M. to 10 A.M.: Hot water is provided by LPG burners.

The SWH system design is developed based on TRNSYS simulation results to meet the hot-water requirements of the hospital from 10 A.M. to 10 P.M. The storage tank is expected to store and deliver hot water during this period.

8   Result of System Performance

SWH installation was finalized in August 2011, and the system has been in operation uninterrupted since. A monitoring system equipped with a weather station and two pyranometers records the presence of solar radiation at the solar panel installation site. The monitoring system records water flow rate, temperature, and pressure at each block of the "energy harvest loop" and "energy transfer loop" along with the temperature of the hot-water storage tank.

Two 2 kW pumps circulate water in both the energy harvest and energy collection loops. The energy harvest loop operates daily from 10 A.M. until 8 P.M. The control system monitors the temperature of the energy harvest loop's water output and compares it with the temperature of the water in the storage tank. The circulation pump of the energy transfer loop operates once the output temperature of the energy harvest loop is 5 °C higher than the temperature of the water in the storage tank.

8.1   Operating Temperature

The installation of a large-scale SWH system with a U-pipe was finalized over 5 years ago, and the system has been in operation since. Hot-water consumption in the hospital was monitored starting a year prior to the SWH installation and continued for over 2 years after completion.

A monitoring system records the data from all sensors every 3 min. Figure 12 shows the annual temperature fluctuation in the hot-water storage tank. As expected, storage tank temperature does not exceed 90 °C and rarely drops below 35 °C.

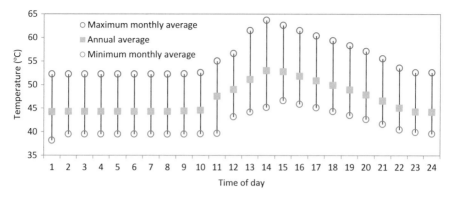

Fig. 11 Annual average, maximum monthly average, and minimum monthly average daily temperature variation in the storage tank

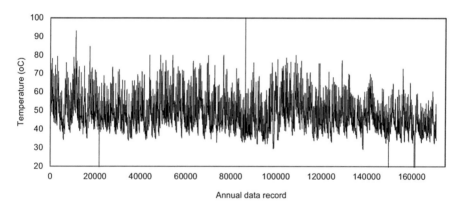

Fig. 12 Annual hot-water storage tank temperature

Daily temperature variation in the storage tank is determined based on Fig. 11. The annual average is listed beside the highest and lowest monthly average daily temperatures of the storage tank in Fig. 12. Hot-water temperature typically remains higher than 45 °C, which is sufficient for the majority of hot-water consumers in the hospital.

The energy absorbed by the solar collectors is transferred to the storage tank by the energy harvest and energy transfer loops. The storage tank discharges hot water to the hospital's hot-water pipeline. Figure 12 displays the measured annual average daily temperature of hot water discharged to the system. "Maximum monthly average" and "minimum monthly average" in Fig. 13, respectively, refer to the hourly maximum and minimum monthly average temperatures observed during 1 year of data collection.

Table 2 and Fig. 14 present the experimental results. Storage tank temperature denotes average temperature inside the storage tank from 10 A.M. to 10

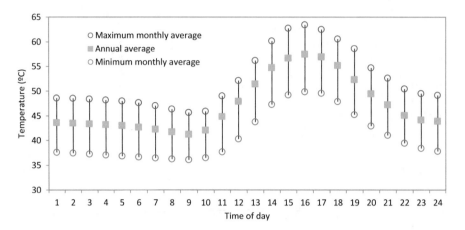

Fig. 13 Annual average daily hot-water temperature

Table 2 Annual system energy and exergy performance evaluation

Parameters	Jan	Feb	Mar	Apr	May	June	July	Aug	Sep	Oct	Nov	Dec	Ave.
Energy efficiency (%)	63.2	62.8	61.1	63.0	64.1	57.7	55.3	59.2	56.2	62.4	58.0	56.2	59.9
Exergy efficiency (%)	5.7	5.6	4.7	5.7	5.8	4.6	4.4	5.5	4.5	5.2	4.1	3.9	5.0

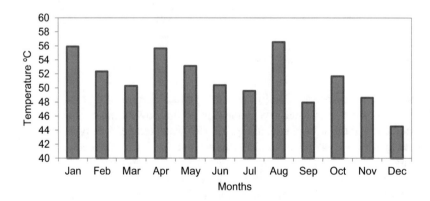

Fig. 14 Monthly average temperature in the storage tank

P.M. However, energy and exergy efficiencies are related to energy harvest (throughout the day) and energy supply periods (10 A.M. to 10 P.M.). Energy efficiency depends on the first law of thermodynamics, whereas exergy efficiency relies on the second law of thermodynamics.

8.2 Operational Efficiency

The hybrid solar–LPG water heater was monitored for over 2 years after the completion of the project. Data collected from January 2012 until December 2012 was used to calculate system energy and exergy performances.

The solar panels are surrounded by hospital buildings at HUKM, which partially shade the solar panels in the late afternoon. A simulation has been carried out to calculate the shading effect prior to project execution. The solar radiation and shading effects have been measured upon completion of project. The simulation results are shown alongside the measured data in Fig. 15. The difference between simulation result and measured data is shown in Fig. 16.

System energy and exergy efficiencies vary throughout the year because of variations in ambient temperature, availability of solar radiation, and hospital occupancy. Annual average energy efficiency is 59.9%, annual exergy efficiency is 5.0%, and annual average storage tank temperature is 51.4 °C, as shown in Fig. 13 and Table 2.

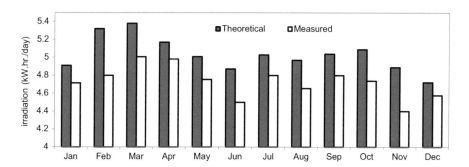

Fig. 15 Theoretical and measured monthly average irradiations

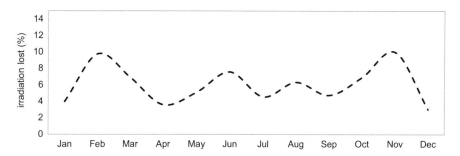

Fig. 16 Percentage of annual irradiation lost as a result of shading

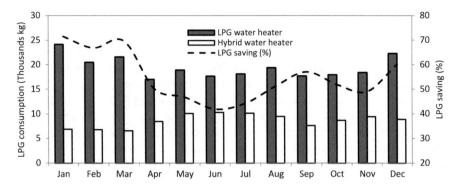

Fig. 17 Monthly LPG consumption in LPG and hybrid modes

8.3 Energy Saving

HUKM has used LPG in water heating for many years. The hospital fully implemented the SWH system for water heating in August 2011. SWH system capacity depends on the number of solar panels, which was limited to 144 because the allocated installation area was small. The hybrid solar LPG system provides all of the hot water required by the hospital. A gas meter has been recording boiler LPG consumption since August 2010. Figure 17 shows the monthly LPG consumption of the hospital from August 2010 until August 2011 and from August 2011 to August 2012 in LPG and solar–LPG hybrid modes, respectively.

Annual average LPG saved during the monitoring period of August 2010 to August 2012 is 51%.

8.4 CO₂ Emission Reduction and Carbon Credit Earned

The simplified form of arbitrary hydrocarbon combustion under stoichiometric conditions is written as (Turns 2000):

$$C_xH_y + a(O_2 + 3.76N_2) \xrightarrow{\text{yields}} bCO_2 + dH_2O + fO_2 + 3.76aN_2 \quad (12)$$

Propane (C_3H_8) and butane (C_4H_{10}) are the main components of LPG. Burning C_3H_8 and C_4H_{10} generates CO_2 and H_2O along with considerable thermal energy (Eq. 1).

$$\begin{cases} 2C_4H_{10} + 13O_2 \xrightarrow{\text{yields}} 8CO_2 + 10H_2O + \text{heat} \\ C_3H_8 + 5O_2 \xrightarrow{\text{yields}} 3CO_2 + 4H_2O + \text{heat} \end{cases} \quad (13)$$

Several varieties of LPG exist worldwide. The LPG mixture in Malaysia is approximately 70% C_3H_8 and 30% C_4H_{10}. Burning 1 kg of Malaysian LPG produces 3.02 kg CO_2.

$$\begin{cases} 70\%\left(1 \ \text{moleC}_4H_{10} + 6.5 \ \text{moleO}_2 \xrightarrow{\text{yields}} 4 \ \text{moleCO}_2 + 5 \ \text{moleH}_2O\right) \\ 30\%\left(1 \ \text{moleC}_3H_8 + 5\text{moleO}_2 \xrightarrow{\text{yields}} 3 \ \text{moleCO}_2 + 2 \ \text{moleH}_2O\right) \end{cases} \quad (14)$$

C_4H_{10}, C_3H_8, and CO_2 have molecular weights of 44.10, 58.12, and 44.01 g/mol, respectively. Substituting the molecular weights of C_4H_{10}, C_3H_8, and CO_2 into Eq. 3 results in:

$$\begin{cases} 70\%\left(58.12 \ \text{kg} \ C_4H_{10} + 104 \ \text{kg} \ O_2 \xrightarrow{\text{yields}} 4 \times 44.01 \ \text{kg} \ CO_2\right) \\ 30\%\left(44.10 \ \text{kg} \ C_3H_8 + 80 \ \text{kg} \ O_2 \xrightarrow{\text{yields}} 3 \times 44.01 \ \text{kg} \ CO_2\right) \end{cases} \quad (15)$$

One mole of LPG is approximately 53.91 kg, and burning it at a stoichiometric air–fuel ratio generates 162.88 kg CO_2.

$$1 \ \text{mole LPG} + 6.05 \ \text{mole} \ O_2 \xrightarrow{\text{yields}} 3.7 \ \text{mole} \ CO_2 + 4.1 \ \text{mole} \ H_2O \quad (16)$$

$$53.91 \ \text{kg LPG} + 96.8 \ \text{kg} \ O_2 \xrightarrow{\text{yields}} 162.88 \ \text{kg} \ CO_2 + 73.86 \ \text{kg} \ H_2O \quad (17)$$

Burning 1 kg of LPG produces over 3.02 kg CO_2. CO_2 emission reduction is calculated based on the gas meter reading. Annual LPG saved reaches 120,482 kg, resulting in a 361,446 kg reduction in CO_2 emission. The international carbon price is between 13 \$/tCO_2 and 16 \$/t CO_2 (Rajoria et al. 2013) that results to over 5700 \$ carbon credit per year for two LPG boilers.

9 Conclusion

A large-scale SWH was designed and installed at HUKM, which houses 1000 beds and is a modern, environmentally friendly training hospital in Malaysia. The SWH consists of 144 U-pipe type evacuated solar panels and is integrated with the existing LPG burner to produce a hybrid solar–LPG water heater. The entire system was installed on a ground-level public area of the hospital to increase societal awareness regarding the benefits and feasibility of large-scale SWHs. The hybrid system was monitored for over 2 years after the completion of the project. Annual LPG saved reached 51% in 2012, resulting in a simultaneous annual reduction of 361,556 kg CO_2 emission and over 5700 \$ carbon credit. A total 60% of LPG saving were expected based on the simulation result; thus, the 9% difference between the theoretical and experimental results is attributed to shading from surrounding buildings and the neglect of thermal loss in the hot-water pipeline in the simulation of the TRNSYS software. Experimental monthly average energy efficiency varies from 55.3% to 64.1%, and exergy efficiency varies from 3.9% to 5.8%. Annual average system energy and exergy efficiencies are 59.9% and 5.0%, respectively. The large-scale SWH has been integrated successfully into an existing public building, and the results affirm the environmental benefits of developing numerous large-scale SWHs for residential, commercial, and industrial water heating applications in Malaysia.

References

Ayompe LM et al (2011) Validated TRNSYS model for forced circulation solar water heating systems with flat plate and heat pipe evacuated tube collectors. Appl Therm Eng 31 (8–9):1536–1542

Borel L, Favrat D (2010) Thermodynamics and Energy Systems Analysis From Energy To Exergy. Engineering sciences. Mechanical engineering. CRC. Taylor and Francis Group, LLC, Boca Raton, FL

Calise F et al (2013) Dynamic simulation of a novel high-temperature solar trigeneration system based on concentrating photovoltaic/thermal collectors. Energy 61(0):72–86

Carrillo Andrés A, Cejudo López JM (2002) TRNSYS model of a thermosiphon solar domestic water heater with a horizontal store and mantle heat exchanger. Sol Energy 72(2):89–98

Chow TT et al (2009) Energy and exergy analysis of photovoltaic–thermal collector with and without glass cover. Appl Energy 86(3):310–316

Duffie JA, Beckman WA (1980) Solar Engineering of Thermal Processes, 2nd edn. John Wiley & Sons Inc, New York, NY

Elmosbahi MS et al (2012) An experimental investigation on the gravity assisted solar heat pipe under the climatic conditions of Tunisia. Energy Convers Manag 64(0):594–605

Fisch MN, Guigas M, Dalenbäck JO (1998) A review of large-scale solar heating systems in europe. Sol Energy 63(6):355–366

Foster R, Ghassemi M, Cota A (2009) Solar energy: renewable energy and the environment. CRC Press, Taylor and Francis Group, LLC, Boca Raton, FL

Fujisawa T, Tani T (1997) Annual exergy evaluation on photovoltaic-thermal hybrid collector. Sol Energy Mater Sol Cells 47(1–4):135–148

Gang P et al (2012) Experimental study and exergetic analysis of a CPC-type solar water heater system using higher-temperature circulation in winter. Sol Energy 86(5):1280–1286

googlemap. Available from: https://maps.google.com.my.

Hepbasli A (2008) A key review on exergetic analysis and assessment of renewable energy resources for a sustainable future. Renew Sustain Energy Rev 12(3):593–661

Islam MR, Sumathy K, Ullah Khan S (2013) Solar water heating systems and their market trends. Renew Sust Energ Rev 17:1–25

itouchmap.com. Available from: http://itouchmap.com/latlong.html

Ji J et al (2007) A sensitivity study of a hybrid photovoltaic/thermal water-heating system with natural circulation. Appl Energy 84(2):222–237

Kim Y, Seo T (2007) Thermal performances comparisons of the glass evacuated tube solar collectors with shapes of absorber tube. Renew Energy 32(5):772–795

Klein SA (1979) TRNSYS, a transient system simulation program: Solar Energy Laborataory, Laboratory, U.O.W.– M.S.E. University of Wisconsin--Madison

Liang R et al (2011) Theoretical and experimental investigation of the filled-type evacuated tube solar collector with U tube. Sol Energy 85(9):1735–1744

Liang R et al (2013) Performance analysis of a new-design filled-type solar collector with double U-tubes. Energ Buildings 57(0):220–226

Ma L et al (2010) Thermal performance analysis of the glass evacuated tube solar collector with U-tube. Build Environ 45(9):1959–1967

Mekhilef S, Saidur R, Safari A (2011) A review on solar energy use in industries. Renew Sust Energ Rev 15(4):1777–1790

Mekhilef S et al (2012) Solar energy in Malaysia: Current state and prospects. Renew Sust Energ Rev 16(1):386–396

Nayak S, Tiwari GN (2008) Energy and exergy analysis of photovoltaic/thermal integrated with a solar greenhouse. Energ Buildings 40(11):2015–2021

Petela R (2003) Exergy of undiluted thermal radiation. Sol Energy 74(6):469–488

Rajoria CS, Agrawal S, Tiwari GN (2013) Exergetic and enviroeconomic analysis of novel hybrid PVT array. Sol Energy 88(0):110–119

Sarhaddi et al (2010) Exergetic performance assessment of a solar photovoltaic thermal (PV/T) air collector. Energ Buildings 42(11):2184–2199

Shahsavar A, Ameri M (2010) Experimental investigation and modeling of a direct-coupled PV/T air collector. Sol Energy 84(11):1938–1958

Sopian K, Othman MYH (1992) Estimates of monthly average daily global solar radiation in Malaysia. Renew Energy 2(3):319–325

Turns SR (2000) An introduction to combustion, 2nd edn. McGraw Hill, Boston

Xiaowu W, Ben H (2005) Exergy analysis of domestic-scale solar water heaters. Renew Sust Energ Rev 9(6):638–645

Zain-Ahmed A et al (2002) The availability of daylight from tropical skies—a case study of Malaysia. Renew Energy 25(1):21–30

Enhancement of Biogas Production from Anaerobic Digestion of Disintegrated Sludge: A Techno-Economic Assessment for Sludge Management of Wastewater Treatment Plants in Vietnam

Khac-Uan Do, Hidenori Harada, and Izuru Saizen

1 Introduction

1.1 Waste Activated Sludge

The activated sludge process is the most widely used biological wastewater treatment for both domestic and industrial plants in the world (Grady et al. 1999; Tchobanoglous et al. 2003). The basic function of a wastewater biological treatment process is to convert organics to carbon dioxide, water and bacterial cells. The cells can then be separated from the purified water and disposed of in a concentrated form called excess sludge. It must be realized that the excess sludge generated from the biological treatment process is a secondary solid waste that must be disposed of in a safe and cost-effective way (Liu 2003). The increased excess sludge production is generating a real challenge in the field of environmental engineering technology (Velho et al. 2016). For example, a quasi-exponential growth of excess sludge production in the USA has been reported (Seiple et al. 2017). In the European Union (EU), some 6 million tons of sludge were generated annually in 1998. By the year 2005, within the EU, about 10 million dry tons per year were generated (Christodoulou and Stamatelatou 2016). Vietnam is also facing the challenge of trying to keep pace with the increasing environmental pollution associated with rapid urbanization and industrialization. Over the past 20 years, Vietnam has made considerable effort to develop urban sanitation policies, legislations and regulations and to invest in urban sanitation including wastewater treatment plants (WWTPs)

K.-U. Do (✉)
School of Environmental Science and Technology, Hanoi University of Science
and Technology, Hanoi, Vietnam
e-mail: uan.dokhac@hust.edu.vn

H. Harada · I. Saizen
Graduate School of Global Environmental Studies, Kyoto University, Kyoto, Japan

© Springer International Publishing AG, part of Springer Nature 2018   129
H.-Y. Chan, K. Sopian (eds.), *Renewable Energy in Developing Countries*,
Green Energy and Technology, https://doi.org/10.1007/978-3-319-89809-4_9

(WorldBank 2013). Most of the domestic wastewater in urban areas is not centrally treated but only treated by household's septic tank and discharged directly into the environment, such as rivers, lakes and streams. Only a few big cities have centralized wastewater treatment plants (Bao et al. 2013). Currently, in Vietnam 35 urban WWTPs had been constructed in Hanoi, Ho Chi Minh City and Da Nang, Quang Ninh, Vinh, Dong Hoi, Quy Nhon, Nha Trang, Da Lat and Buon Ma Thuot cities with a total capacity of 850,000 m^3/day. Some 40 new WWTPs are in the design or construction phase with a capacity of 1,600,000 m^3/day (Nga 2017). Industrial parks in Vietnam have developed rapidly since 1986, and there are 289 industrial parks throughout the country (Thuy et al. 2016). Until now there have been about 90% of the industrial parks which have WWTPs that are operated or under constructed (Pham et al. 2016). To date, in Vietnam the wastewater treatment technology has been focused on the use of some form of activated sludge secondary treatment technology, such as conventional activated sludge (CAS), anaerobic-anoxic-aerobic (A2O), oxidation ditch (OD) and sequencing batch reactor (SBR) technologies (WorldBank 2013, 2014). During operation, a lot of sludge has been discharged. The sludge disposal has been a significant challenge and attracted great attention in both academic and engineering fields. It should be noted that the cost of the excess sludge treatment and disposal can account for 30–40% of the capital cost and about 50–60% of the operating cost of many wastewater treatment facilities (Nowak 2006; Wang et al. 2013). Moreover, the conventional disposal methods such as landfill or ocean dumping may cause secondary pollution problems and are strictly regulated in many countries (Oh et al. 2007). In Europe, sea disposal has not been used since the end of 1998. In 2000, landfill of dewatered sludge became illegal in the Netherlands. Landfilling of sludge has been totally forbidden in Europe since 2015 (Christodoulou and Stamatelatou 2016). However, in Vietnam, sludge discharged from wastewater treatment plants were mostly disposed by the conventional method such as landfill (Uan et al. 2016). Therefore, reduction of sludge volume before disposal is very important for sludge transportation and management. It should be noted that the anaerobic digestion could be used as a common method of sludge stabilization (Appels et al. 2008). Besides, anaerobic digestion is a burgeoning technology and has lately captivated much attention owing to the need for sustainable energy production (Appels et al. 2013; Browne et al. 2014; Kavitha et al. 2015a, b). Therefore biogas produced through anaerobic degradation of sludge in wastewater treatment plants has gained much more attention as it is a renewable energy resource (Appels et al. 2008; Yu et al. 2016).

1.2 Anaerobic Digestion of Waste Activated Sludge

Basically, sludge was thickened before introducing to anaerobic degradation process in order to reduce the sludge volume. Besides, hydraulic retention time is identical to solid retention time, leading to a larger volume of anaerobic digester (Wang et al. 2013). Anaerobic digesters have retention times in the range of 20–30 days, and approximately only half of the organic material fed to anaerobic digestion could be degraded and subsequently transformed to methane (Uma Rani et al. 2014). In the anaerobic degradation systems, *Proteobacteria and Bacteroidetes* were the two most predominant phyla in the digested sludge samples (Chouari et al. 2005). *Proteobacteria* are able to degrade a wide range of macromolecules (Chouari et al. 2005; Costa et al. 2017). Bacteroidetes, known to be proteolytic bacteria, are involved in protein degradation and able to ferment amino acids to acetate (Riviere et al. 2009; Świątczak et al. 2017; Yang et al. 2014). Analysis of microbial community structure suggested *Bacteroidetes* and *Firmicutes* were the dominant species when ozone pretreatment was applied (Li et al. 2017b). However, conventional anaerobic degradation process needs substantial improvements, especially for the treatment of sludge with low solids content and poor anaerobic biodegradability (Appels et al. 2008; Carrère et al. 2010; Jimenez et al. 2014; Jin et al. 2004). Besides, recently, anaerobic degradation of sludge has been enhanced by using the submerged anaerobic membrane bioreactor (Baêta et al. 2016; Yu et al. 2014, 2016). Co-digestion of wastewater sludge and food waste has been also applied in full scale anaerobic digesters for biosolids management and biogas generation (Amha et al. 2017; Fitamo et al. 2017; Nghiem et al. 2017). The anaerobic degradation process is achieved through several stages: hydrolysis, acidogenesis, and methanogenesis. For sludge degradation, the rate-limiting step is the hydrolysis. It should be noted that the initial phase of anaerobic degradation, hydrolysis, is considered to be the rate-restricting phase (Chen et al. 2013). Hence, sludge disintegration has to be done prior to the anaerobic degradation in order to increase the hydrolysis rate.

1.3   Sludge Disintegration Methods

It should be noted that in anaerobic degradation process, sludge hydrolysis leads to the rupture of cell walls and the release of extracellular polymeric substances, which provides soluble organic substrates, such as dissolved organic matter, for acidogenic microorganisms (Appels et al. 2008). It was observed that pretreatment has to be done prior to anaerobic degradation to enhance the biogas generation. The most attractive way to enhance anaerobic degradation performance is by pretreating the waste to convert insoluble organic polymers into soluble components (monomers) (Nazari et al. 2017). By doing this it is possible to enhance hydrolysis rate, which subsequently increase the biogas generation, as well as reduce the digestion time and the amount of final residuals (Banu and Kavitha 2017). There are various sludge disintegration techniques attracted attentions as promising alternatives to reduce sludge production. Sludge disintegration techniques have been reported to enhance the biodegradability of sludge. Sludge disintegration methods reported in the literature include both physical methods such as ultrasound (Han et al. 2013; Hirooka et al. 2009), ball mill, and homogenizer treatments and chemical methods such as ozone (Muz et al. 2014; Sallanko and Okkonen 2009; Vlyssides and Karlis 2004; Aquino and Pires 2016), acid (Velho et al. 2016) and alkali treatments (Do et al.

Table 1 Sludge disintegration processes

Methods	Disintegration processes			References
Mechanical	Stirred ball mill	High-pressure homogenizer	Ultrasound	Øegaard (2004), Appels et al. (2013)
Physical	Thermal treatment	Osmotic shock	High-yield pulse	Atay and Akbal (2016), Banu and Kavitha (2017)
Chemical	Acid or base hydrolysis	Oxidation with ozone	Oxidation with H_2O_2/O_2/Fentons reagent	Fang et al. (2014), Chen et al. (2013)
Biological	Enzymatic lysis	Autolysis	Cell lysis-cryptic growth	Chouari et al. (2005), Velho et al. (2016)

2009; Oh et al. 2007). Besides, thermal treatment (Higgins et al. 2017; Kim et al. 2016), cell lysis-cryptic growth (Romero et al. 2013) and enzyme treatment (Ohsaka 2005; Song et al. 2013) have also been tested. Recently, a novel and energy-efficient radio frequency pretreatment system has been developed for anaerobic digestion of municipal sludge (Barrios et al. 2017; Hosseini Koupaie et al. 2017). The aim of all pretreatments is to disintegrate the sludge flocs, disrupting the cell wall, thus releasing and solubilizing intracellular material into the liquid phase. Several researchers have reported affirmative synergistic upshots of the combined pretreatment methods on subsequent anaerobic digestibility (Feki et al. 2015; Kavitha et al. 2015b; Li et al. 2017a, b; Pilli et al. 2015; Şahinkaya et al. 2012; Tyagi and Lo 2012; Wang et al. 2016b; Zhao et al. 2017; Zhen et al. 2014). In this process, the sludge disintegration enhances transformation of particulate organic compounds into more readily biodegradable substances and subsequently accelerates the process of anaerobic methane production (Do et al. 2009; Hirooka et al. 2009; Li et al. 2017c; Liu 2003; Nowak 2006; Øegaard 2004; Oh et al. 2007; Pérez-Elvira et al. 2006; Wang et al. 2016a; Yu et al. 2016). The disintegration processes are based on mechanical, thermal, chemical or thermochemical and biological techniques shown in Table 1.

Basically, mechanical sludge disintegration methods are generally based on the disruption of microbial cell walls by shear stresses. Cells are disrupted when the external pressure exceeds the cell internal pressure (Banu and Kavitha 2017). Mechanical disruption of sludge has gained acceptance due to its various successful industrial scale applications. The high-energy levels were most probably the reason why the application of mechanical disruption methods is still limited. Heat treatment results in the breakdown of the gel structure of the sludge and the release of intracellular bound water. Thermal hydrolysis involves heating of the sludge. Increased temperature had a major positive effect on the yields of soluble COD (Atay and Akbal 2016). Apparently, the origin of the sludge is of primary importance for the final solubilization to be reached with thermal hydrolysis. In chemical and/or thermochemical hydrolysis techniques, an acid or base is added to solubilize the sludge cells (Ødegaard et al. 2002). Whereas for thermal destruction methods high temperatures are required to achieve acceptable results, the thermochemical treatments are often carried out at lower or ambient temperatures. With respect to alkaline pretreatments, variable results have been found. An additional advantage of alkali instead of acid is that it is readily compatible with subsequent biological treatment (Wang et al. 2016b). Biological hydrolysis can be considered as a partial anaerobic sludge digestion. The biochemical sludge disintegration processes are based on enzyme activity that is either produced within the system. Biological hydrolysis is an easy and inexpensive method for the in situ production of a readily degradable carbon source for nutrient removal (Velho et al. 2016). An additional advantage is that less sludge is produced, compared with a system with external carbon addition.

1.4 Brief Economic Assessment of Sludge Disintegration Methods

The strategies for sludge disintegration should be evaluated and chosen for practical application using costs analysis and assessment of environmental impact. Economic savings from the reduced costs of treatment and disposal of biomass improved operational efficiencies and reduced environmental burden with lower disposal requirements (Atay and Akbal 2016). Other economic, operational and environmental costs may be incurred and these must be considered. The environmental impact, e.g. odour problems, should be assessed (Carrère et al. 2010). The performance of some disintegration methods can be compared with each other using the specific energy, which is defined as the amount of energy that stresses a certain amount of sludge. Müller (2001) has carried out a comparison in terms of five aspects (i.e. rate of sludge degradation, degree of sludge degradation, bacteria disinfection, influence on the dewatering results and odour generation). The author found that the mechanical methods contributed an excellent role in sludge degradation rate. However, the chemical methods using ozone could give the highest degree of sludge degradation compared with other methods. Among the summarized methods, the thermal methods could have a strong effect on bacterial disinfection. It seems that the odour generation was not affected by the sludge disintegration methods. It should be due to the sludge disintegration processes that were mostly carried out in the closed reactors.

It should be noted that the mechanical disintegration has been investigated primarily on laboratory to pilot scale. Problems encountered were the heating of the cell suspension because of the high shear stresses the sludge cells are being subjected to. Moreover, mechanical disintegration often appears to require high capital equipment and is energy intensive (Han et al. 2013). On the other hand, thermal and thermochemical treatments require high temperatures and high pressures to achieve acceptable results. Not only is equipment needed to raise the temperature and the pressure, also expensive construction materials are required in order to prevent corrosion problems (Carrère et al. 2010). Furthermore, odour problems can be encountered in thermal hydrolysis techniques. Most authors

Table 2 A comparison of several sludge pretreatments

Parameters	Relatively high			Relatively low	References
Energy demand	Lysate centrifuge	Stirred ball milling	Sonication	Ozonation	Müller et al. (2004)
Sludge degradation	Ozonation	Stirred ball milling	Sonication	Lysate centrifuge	Carrère et al. (2010)
Polymer demand for dewatering	Ozonation	Sonication	Stirred ball milling	Lysate centrifuge	Fang et al. (2014)
Soluble COD and ammonia concentrations in supernatant after dewatering	Ozonation	Stirred ball milling	Lysate centrifuge	Sonication	Yang et al. (2013)

mention that acidic or basic conditions should be applied in combination with elevated temperatures, thereby creating quite aggressive reaction conditions (Yang et al. 2013). Moreover, raising or lowering the pH requires the addition of chemicals which increase the ionic strength of the sludge (Fang et al. 2014). If the hydrolysate is used in biological applications, e.g. anaerobic digestion or nutrient removal, subsequent neutralization is required, which again implies the addition of chemicals. In addition, due to high costs caused from ozone production, it is important to decrease the amounts of ozone required for sludge reduction.

It has been reported that one of the most significant inputs, environmentally and financially, is energy. While the cost of treatment may be disposal driven, in energy terms, energy utilized should hopefully match the energy produced by increases in biogas production. The energy input depends heavily on the method and may be a function of sludge composition, operating and ambient conditions and equipment used, among others (Carrère et al. 2010). A comparison of several sludge pretreatments such as stirred ball milling, ozonation, lysate centrifugation and sonication could be classified according to the aspects below (Table 2).

The evaluation of energy balance and cost assessment should be used in the conventional data such as industrial power price of 0.23 USD/kWh, NaOH price of 345.6 USD/ton and sludge treatment and disposal cost of 441.2 USD/ton TS. Meanwhile, energy stored in the increased methane volume as a result of pretreatment reached 378.15 kWh and 751.08 kWh, respectively (Kavitha et al. 2016). An energy balance and cost assessment of the thermo-chemo-sonic disintegrated sludges were performed by Kavitha et al. (2015b). In the study, the three alkalis (NaOH, KOH and $Ca(OH)_2$) were used in the thermo-chemo-sonic disintegrated sludge process. The energy balance and cost evaluation were based on the energy content of the biogas produced from both samples. The energy utilized for the mechanical stirring and pumping of the sludge was taken into account. At the same time, the energy content of the biogas for all three samples was calculated to be 1213 kWh, 1043 kWh and 927 kWh, respectively. Moreover, to evaluate the economic viability of the disintegration process, the present study took into account

Table 3 Cost comparison of sludge disintegration methods

No.	Sludge disintegration methods	Cost	References
1	Disperser induced microwave pretreatment	A net profit of 104.8 USD/t TS was achieved for disperser induced microwave pretreated sludge	Kavitha et al. (2016)
2	Electric + alkaline	Highest methane yield of combined electrical alkali pretreatment spurred a 20.3% increase	Zhen et al. (2014)
3	Alkaline + pressure homogenization	An enhancement of 247 mL/g VS methane production and of 43.5% VS removal was achieved	Fang et al. (2014)
4	Microwave + alkaline	Nearly 83.39 USD/t TS was gained from the combined pretreatment of microwave and alkaline with a 13% increase in biodegradability and a 28% VS reduction	Yang et al. (2013)
5	Ultrasonic + alkaline	Total revenue of 21.54 USD/t TS when the VS destruction rate is increased to 46% and biogas price reached 10 USD/GJ	Park et al. (2012)

an estimation of the operational cost (including consumable chemicals) and the decreased amount of the solids to be disposed. It was noted that a positive net profit was achieved in all three samples. The net profit of all three sludges (NaOH, KOH and $Ca(OH)_2$ was calculated to be 42.6 USD, 20.6 USD and 4 USD, respectively. The presently attained net profit was found to be comparatively higher than those obtained in other studies in which Houtmeyers et al. (2014) achieved 2.48 USD and 3.02 USD as net costs for ultrasonic and microwave pretreatment. It should be noted that in low-temperature thermal treatment, it would take from several hours to a few days to achieve the maximal disintegration effect (Şahinkaya et al. 2012). This strongly implies that it was possible to achieve 20% solubilization only in this combined novel process with lesser energy consumption. Table 3 shows a cost comparison of sludge of some combined sludge disintegration methods.

Carrère et al. (2010) have analysed the energy consumption for the various sludge disintegration methods, including non-mesophilic, non-thermophilic, biological, thermal hydrolysis, sonication, ball milling and high pressure. The energy consumption in those methods was mainly electrical and thermal. Electrical requirements are mainly feed and mixing and are approximately $0.1–0.2$ kWh/m^3 days. The analysis also assumes a hydraulic retention time of 20 days for mesophilic or 15 days for thermophilic. It should be noted that the heating requirements are thermal capacity plus about 10% losses in mesophilic or 20% in thermophilic (Greenfield and Batstone 2005). The authors found that the electrical consumption was varied from 0.03 to 0.04 kWh/kg VS (volatile solids) for the non-mesophilic, non-thermophilic, biological and thermal hydrolysis methods. However, it was much higher in case of sonication (0.37 kWh/kg VS), ball milling (1.04 kWh/kg VS) and high pressure (0.33 kWh/kg VS). Basically, the thermal consumed in most of the methods was in the range of $0.5–1.0$ kWh/kg VS, except thermal hydrolysis (2.0 kWh/kg VS) (Carrère et al. 2010).

1.5 General Objectives and Scopes of This Work

It should be noted that among the sludge disintegration methods, chemical hydrolysis using alkaline was the most efficient for inducing cell lysis (Do et al. 2009; Banu and Kavitha 2017). The chemical-combined activated sludge processes would be more efficient for sludge disintegration. The chemical assisted sludge disintegration processes have advantages of easy control, stable performance and high operation flexibility. The minimum effective sludge disintegration index for anaerobic digestion was reported to be 25% (Gayathri et al. 2015; Zhang et al. 2008). However, achieving solubilization in excess of 18% by thermochemical pretreatment was not cost effective (Jang and Ahn 2013) but results in the loss of organics (Chiang et al. 2012). Sonication, a cavitational process, was used by legions of researchers to achieve high degree of solubilization (40–50%). However, the practical applicability of sonic pretreatment was constrained because of its high energy cost (Şahinkaya et al. 2012; Zhang et al. 2008). Thus, in order to overcome the high energy requirement, the sonic pretreatment can be combined with other pretreatments to achieve the desirable solubilization with less energy consumption. Besides, the alkaline treatment is known to be relatively cheap resulting in a significant decrease of the total treatment cost. Therefore, in this study, sludge disintegration using alkalis such as sodium hydroxide (NaOH) and calcium hydroxide ($Ca(OH)_2$) was applied to sludge taken from Yen So and Kim Lien wastewater treatment plants in Hanoi (Vietnam). In particular, the study aimed to examine (i) the sludge solubilization and sludge dewatering ability during treatment, (ii) the change of the biodegradable matter and the particle size distribution and (iii) a brief economic evaluation of the alkali sludge disintegration and (iv) to test the biodegradability of sludge after alkali digestion. The digested sludge is subsequently treated anaerobically to generate biogas, and then the biogas can give long-term economic benefits.

2 Materials and Methods

2.1 Sludge Sources and Collection

In this study, waste activated sludge was collected from Yen So and Kim Lien wastewater treatment plants in Hanoi (Vietnam) (Fig. 1). These sludge samples were taken from oxic tank, SBR tank and sludge storage tank. The Yen So wastewater treatment plant works on the principle of SBR technology with a design capacity of 200.000 m^3/day, whereas the Kim Lien wastewater treatment plants work on A^2O technology with the capacity of 3500 m^3/day. The sludge samples were collected weekly and stored in refrigerator (4 °C) until use within 24 h. General characteristics of sludge samples are presented in Table 4.

Fig. 1 Sludge samples taken from Kim Lien and Yen So wastewater treatment plants

Table 4 General characteristics of sludge used in this study

Parameters	Kim Lien (storage tank)	Kim Lien (oxic tank)	Yen So (storage tank)	Yen So (SBR tank)
pH	7.1	7.3	7.2	7.2
MLSS, mg/L	4696	1850	9660	4948
MLVSS, mg/L	2728	820	5986	3034
MLVSS/MLSS	0.58	0.44	0.62	0.61
Soluble COD, mg/L	53	41	95	32
Total COD, mg/L	4410	1730	9740	5040

2.2 Experimental Setup and Procedure

Sludge disintegration experiments will be carried out at laboratory temperature using a jar apparatus with six paddle stirrers (Model: JLT6, VELP Scientifica, Europe). Alkalis were added under stirring. Rapid mix took place for 1 min at a speed of 200 rpm, followed by slow mix for 3 h at 30 rpm. During digestion, NaOH and Ca(OH)$_2$ were added to the reactor at various dosages, ranging from 0.2 to 2 g/L (Fig. 2).

The anaerobic digestion experiments were conducted in a lab-scale anaerobic digestion system (Figs. 3 and 4).

The anaerobic reactor is a round-bottomed reactor with a working volume of 5 L. The pretreated sludge was inoculated with an active methanogenic bacterial population for the quick startup of the reactor. Sludge from an anaerobic wastewater treatment using a covered lagoon was selected as the inoculum. A slow mixer was used to keep a completely mixed anaerobic digester. Biogas generation was calculated by the liquid displacement method, in which the difference in the water level in a cylinder linked to the reactors was measured. The displaced liquid is therefore considered to have the same volume as the produced biogas. The produced gas was collected in standardized glass cylinders filled with acidified deionized water to evade losses of CO_2 due to the formation of carbonates.

Fig. 2 Sludge digested in jar tests with different NaOH and Ca(OH)$_2$ dosages. (**a**) NaOH dosages. (**b**) Ca(OH)$_2$ dosages

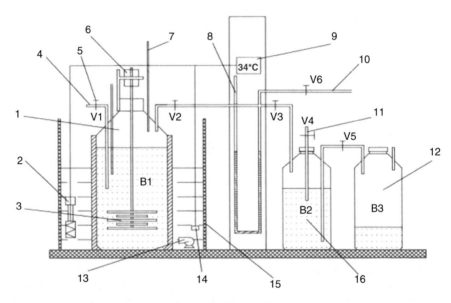

Fig. 3 Setup of a simplified test for measuring the biogas production. Notes: 1 Anaerobic reactor (V = 5 L), 2 Heating in water batch, 3 Mixer, 4 Feeding pipe, 5 Valve, 6 Motor, 7 Thermostart, 8 Pressure pipe, 9 Temperature control, 10 Biogas collection pipe, 11 Water addition pipe, 12 Gas volume bottle, 13 Mixing pump in water batch, 14 Temperature sensor, 15 Water batch cover, 16 Water bottle

2.3   Analytical Methods

Characteristics of sludge (including MLSS, MLVSS, sludge dewatering), and other parameters of the solution such as TP, soluble TP, total COD, soluble COD, BOD, TN, soluble TN, and NO$_3^-$ were measured and analysed in accordance with the Standard Methods described in APHA (2012). Sludge disintegration efficiency was calculated as the ratio of the soluble COD (SCOD, mg/L) increase by the disintegration process to the total COD (TCOD, mg/L) of the sludge before disintegration.

Fig. 4 A photo of the anaerobic digestion system used in this study

$\alpha = \frac{SCOD - SCOD_o}{TCOD - SCOD_o}$ in which $SCOD_o$ (mg/L) is the soluble COD of the sludge before disintegration.

The biogas composition biogas was taken during experimental tests. The compositions of these samples (CH_4, CO_2, H_2S, O_2) were determined using an Optima 7 Biogas Analyzer (MRU Instruments, Inc., Germany).

3   Results and Discussion

3.1   Sludge Source Variation

Sludge used in this study were collected from the Yen So and Kim Lien wastewater treatment plants in Hanoi (Vietnam). It should be noted that the SBR technology with a design capacity of 200,000 m³/day was used in the Yen So wastewater treatment plant, but the Kim Lien wastewater treatment plant used A²O technology with a design capacity of 3500 m³/day. The variation of sludge sample taken from sludge storage tank was presented in Fig. 5. As seen from the figure, the sludge concentration was very different from each other. In Yen So wastewater treatment plant, it was in a range of 4000 to about 10,000 mg/L. In case of Kim Lien wastewater treatment plants, it was around 2500 to over 6000 mg/L.

3.2   Sludge Solubilization

Alkalis added to a cell suspension react with the cell walls in several ways, including the saponification of lipids in the cell walls, which leads to solubilization of the membrane. Disruption of microbial cells then leads to leakage of intracellular material out of the cell. In this study, the collected sludge was tested with different

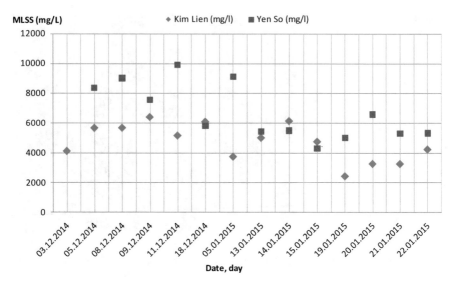

Fig. 5 Variation between MLSS of Yen So and Kim Lien wastewater treatment plants

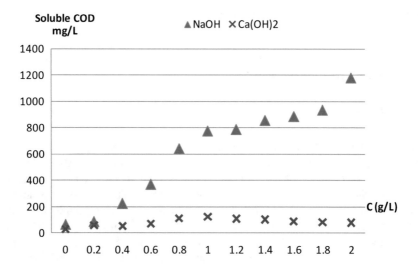

Fig. 6 Variation of soluble COD during digestion

NaOH and $Ca(OH)_2$ dosages. Each 1 L of sludge was added in six glass beakers and put in the jar test system. The first beaker was without added by any alkalis. The mining beakers were added with different NaOH and $Ca(OH)_2$ dosages, from 0.2 to 2 g/L. During digestion, soluble COD will be released out of the sludge. Figure 6 shows the variation of soluble COD during digestion.

It could be seen from Fig. 6 that the soluble COD was increased with the increase of NaOH. However, it was increased very low during adding $Ca(OH)_2$. When alkali agents were added, COD solubilization increased through various reactions such as saponification of uronic acids and acetyl esters, reactions occurring with free carboxylic groups and neutralization of various acids formed from the degradation of particular materials (Liu 2003). During digestion process using $Ca(OH)_2$, calcium ion could bind to the sludge surface through the exopolysaccharide polymer as well as assist in the bioflocculation of the sludge, resulted in reduction of the sludge solubilization (Bruus et al. 1992).

3.3 Sludge Mass Reduction

The reduction of MLSS during digestion was presented in Fig. 7. It was reduced from 6600 mg/L down to about 4800 mg/L when alkali dosage was increased from 1.0 to 1.8 g/L. Besides MLSS reduction, the soluble COD was also increased from less than 100 mg/L to over 1800 mg/L. Alkali treatment is a harsh method. At extremely high pH values of medium, the cell loses its viability, and it cannot maintain an appropriate turgor pressure and disrupts. Alkalis added to the cell suspension react with the cell walls in several ways, including the saponification of lipids in the cell walls, which leads to solubilization of membrane. The high alkali concentrations cause much degradation. Disruption of sludge cells leads to leakage of intracellular material out of the cell (Banu and Kavitha 2017). As a result, the sludge mass was decreased, and the COD was increased. Besides, a relationship between MLSS, MLVSS reduction and soluble COD increasing during digestion for different sludge samples was also measured and presented in Fig. 8. Depending on MLSS concentration, sludge reduction was varied from 10% to 18%.

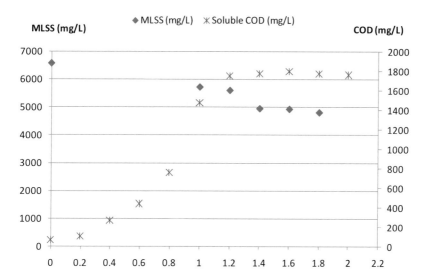

Fig. 7 Relationship between MLSS reduction and soluble COD during digestion

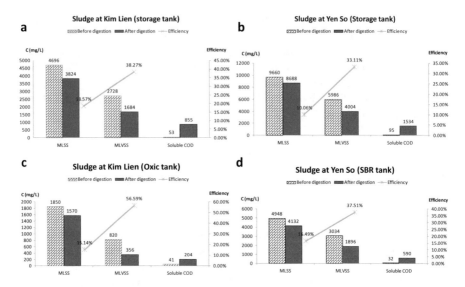

Fig. 8 Relationship between MLSS, MLVSS reduction and soluble COD increasing during digestion for sludge taken from storage tank of Kim Lien WWTP (**a**) and Yen So WWTP (**b**); sludge taken from oxic tank at Kim Lien WWTP (**c**) and from SBR tank in Yen So WWTP (**d**)

3.4 Enhancement of Sludge Dewatering

Sludge dewatering ability was tested by measurement of time to filter. At the same volume of digested sample, the filter time between NaOH and Ca(OH)$_2$ was much different (Fig. 9). The filter time of NaOH digested sample was much higher than Ca(OH)$_2$ digested sample, showing that Ca(OH)$_2$ could help improvement of sludge dewatering ability. One reason for the difficulty in activated sludge dewatering is the presence of extracellular polymer (ECP). ECP is present in varying quantities in sewage sludge, occurring as a highly hydrated capsule surrounding the bacterial cell wall and loose in solution as slime polymers (Houghton et al. 2001). One of the main influences on sludge dewaterability is the particle size distribution (Eriksson and Alm 1991). Flocculation changes the particle size distribution of a sludge, binding small particles together, thereby influencing the sludge dewatering characteristics. ECP can therefore be expected to have an influence on sludge dewaterability through the high level of hydration of the polymer surrounding the bacterial cell and its role in flocculation. Bruus et al. (1992) suggested that cations aid in flocculation by bridging negative sites on ECP which promotes an increase in the floc size, floc density and floc resistance to shear. Divalent cations act as a bridge between negatively charged sites on ECP resulted in improvements of sludge settling and dewatering.

Figures 10 and 11 show the settled volumes and SVI changes of various sludge samples before and after digestion. It seems that after digestion, SVI was improved slightly. The alkali disintegration alters sludge floc properties (such as size, water

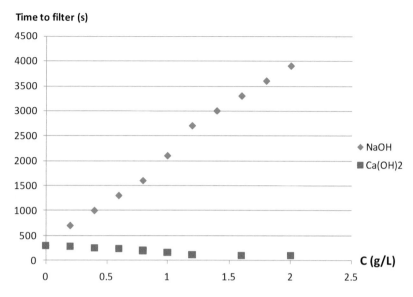

Fig. 9 Variation of time to filter during sludge digestion

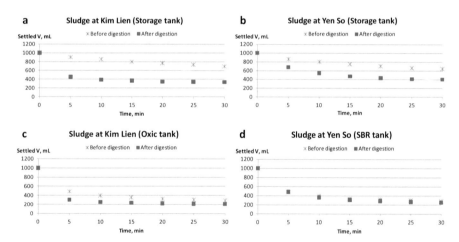

Fig. 10 Settled volumes of various sludge samples before and after digestion sludge taken from storage tank of Kim Lien WWTP (**a**) and Yen So WWTP (**b**); sludge taken from oxic tank at Kim Lien WWTP (**c**) and from SBR tank in Yen So WWTP (**d**)

content, etc.); it also modifies its settling and filtering properties. In particular, the alkali disintegration integrated in the wastewater handling units increases the settling rate and could reduce SVI. The SVI determines the settling capability of sludge. The value of SVI for alkali-disintegrated sludge was found to be in the range of 46–113 mL/g, which was comparatively higher than the others. It is known from the literature that the usage of Na^+ ions deteriorates the dewatering property of sludge, and it thereby demands the addition of sludge-conditioning aids (Banu et al. 2012).

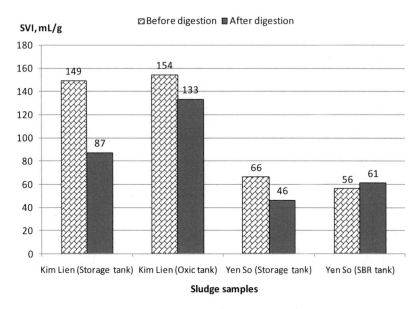

Fig. 11 SVI changes of various sludge samples before and after digestion

Table 5 Biodegradability of solubilized COD of sludge samples after digestion

Parameters	Kim Lien sludge (NaOH digestion)	Kim Lien sludge (Ca(OH)$_2$ digestion)	Yen So sludge (NaOH digestion)	Yen So sludge (Ca(OH)$_2$ digestion)
MLVSS before digestion, mg/L	2728	2728	5986	5986
MLVSS after digestion, mg/L	1684	2486	4004	4004
Soluble COD, mg/L	855	169	1534	207
BOD$_5$, mg/L	706	51	1228	73
BOD$_5$/COD (%)	83	30	80	35

3.5 Biodegradability of Solubilized COD After Alkali Digestion

The biodegradability of solubilized COD after alkali digestion was measured by BOD (Table 5). The results observed that the biodegradability of the solubilized COD by Ca(OH)$_2$ was low, around 30–35%. This unexpected phenomenon is probably due to the release of part of the sludge material by desorption or floc destructuration while no increase in its intrinsic biodegradability occurs. In case of NaOH digestion, the sludge biodegradability enhancement is linearly correlated to COD solubilization up to 80–83%. Due to its enhanced biodegradability, the super-natant of NaOH sludge digestion would be used as a carbon source to support post-

Table 6 Characteristics of sludge samples after digestion

Parameters	Kim Lien (storage tank)	Kim Lien (oxic tank)	Yen So (storage tank)	Yen So (SBR tank)
MLSS, mg/L	3824	1570	8688	4132
MLVSS, mg/L	1684	356	4004	1896
MLVSS/MLSS	0.44	0.23	0.46	0.46
Soluble COD, mg/L	855	204	1534	590
NO_3^-, mg/L	1.5	–	0.8	16.6
NH_4^+, mg/L	4.2	–	48.4	2.5
TN, mg/L	64	–	144	58
TP, mg/L	37.2	–	44.2	16.8
SVI, mL/g	87	133	46	61

denitrification in the biological nitrogen removal. During sludge disintegration, TN, NO_3^-, NH_4^+, TP and PO_4^{3-} releases were also monitored. The results were presented in Table 6. It should be noted that the structure of sludge flocs was significantly dispersed after alkali digestion. The average floc size decreased in the range of about 1–50 μm (Uan et al. 2017). Sludge flocculation capacity decreased with microorganism death.

3.6 Anaerobic Digestion of Digested Sludge

In anaerobic digestion process, the organic residues will be transformed by bacteria into the biogas through four sequential phases, i.e. hydrolysis, acidogenesis, acetogenesis and methanogenesis in anaerobic digestion process. The progression of anaerobic digestion has the prospective of transforming recyclable organics into biogas. On the other hand, in lack of a pretreatment, the biogas production potential of anaerobic digestion is inadequate due to the existence of refractory microbial cell walls and other organic materials present in the sludge. As a result, to enhance the biogas augmentation, pretreatment is believed to be the plausible option. Sludge disintegration was extended to boost bioenergy generation by hastening the hydrolysis pace of anaerobic digestion. To increase the biogas production, assorted pretreatment practices are applied. A comparatively current scientific progression which would probably be capable of formulating anaerobic digestion further was the expansion plus concern of preprocessing of wastes prior to anaerobic digestion to hasten the sludge degradability. Pretreatment enhances liquefaction and its rate, which consequently enhance biogas production. All pretreatments bring about the breakdown of sludge biomass, thus solubilizing and discharging substance present inside the aqueous phase and converting obstinate organic substances into recyclable nature consequently creating substances easily accessible to microbes. These entire pretreatments are exposed to enhance the biogas generation in subsequent anaerobic digestion process. Figure 12 presents the digestion curves obtained in the anaerobic

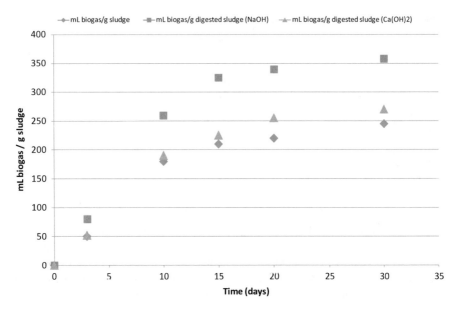

Fig. 12 Biogas production curves in anaerobic digestion tests with sludge with and without alkaline sludge disintegration

digestion tests. It can be observed that the biogas production of the sludge without the alkali disintegration was very slow. As expected, the biogas production of the disintegrated sludge was higher compared to the sludge without the alkali disintegration (358 and 245 mL biogas/g sludge added), representing a 46% increase in the biogas production in case of NaOH disintegration. However, it was only about 10% increase in the biogas production in case of $Ca(OH)_2$ disintegration. The fractioning of the disintegrated sludge showed that the contribution of the enhancement in biogas production and kinetics after the hydrolysis is due to the liquid fraction, remaining the solid fraction difficult to degrade.

The composition of biogas production during anaerobic digestion of the alkaline-disintegrated sludge was sometimes measured by using an Optima 7 Biogas Analyzer (MRU Instruments, Inc., Germany). The results show that the CH_4 composition is high at about 73%. It was tried to burn very well in the lab (Figs. 13 and 14). However, it was clearly that the H_2S concentration in the biogas is too high at about 1989 ppm. It means the biogas must be purified before using such as for burning or for electricity generation.

3.7 Economic Analysis for Sludge Akali Disintegration

As discussed above, the sludge can be disintegrated by mechanical, physical, chemical or biological processes. The main purpose of the disintegration processes

Fig. 13 A photo of the composition of biogas production during anaerobic digestion of the alkali-disintegrated sludge

Fig. 14 A photo of sludge after anaerobic digestion (left) and biogas generation burned (right)

is to improve the anaerobic digestion of the sludge. Therefore, a high degree of disintegration is necessary in order to realize a noticeable acceleration and enhancement of the disintegrated sludge degradation. However, it is not easy to recommend one or the other method to be applied, because of the lack of experience at full scale. A comparison between the capital costs and the operational and maintenance (O&M) costs among treatment processes is needed. Rough cost estimates are between 70 and 150 US$/t TS for capital and O&M costs (Müller 2001). It has been reported that the O&M costs vary considerably among sludge disintegration methods. The O&M costs for sludge disintegration by biological methods were lower compared with others. However, it should be noted that the biological methods required long retention time for the disintegration process, resulted in bigger volume of reactor. The high energy levels were most probably the reason why the application of mechanical disruption methods is still limited. Thermal methods also required high energy for heating. Besides, the thermal treatment time had less impact on sludge solubilization in comparison with temperature (Müller 2001). However, if temperatures are not high enough, several hours to days of heating was required (Zhen et al. 2017). The chemical methods required the chemical costs. During treatment, the corrosion issues should be considered. It is important that the sludge cells can be dissolved by acids of alkalis at low or ambient temperatures. Chemical methods could be realized at low costs if the conditions are appropriate (Uan et al. 2017).

In this study, a preliminary economic analysis was evaluated. In particular, the operating costs for alkali digestion were calculated based on the chemical consumption and TSS concentration or sludge volume. Besides, the costs of dewatering, transportation and landfill should be taken into account. In case of NaOH used for sludge digestion at 1.2 g/L (it means 0.12 kg of NaOH was used for 1 kg of TSS, and the cost of NaOH (99%) of about 0.3 US $/kg), the chemical consumption was estimated about 0.36 US $/kg TSS (about 8,100 VND/kg TSS). In Vietnam, the cost for transportation and treatment of sludge was varied from 0.20 to 0.44 US $/kg TSS (about 4500 to 10,000 VND/kg TSS), depending on the sludge characteristics and services. If the digested sludge is subsequently treated anaerobically to generate biogas, then it can be biogas can give long-term economic benefits.

It should be noted that 1 m^3 of biogas could produce hourly 2.14 kWh of electricity and 2.47 kWh of heat energy (Akbulut 2012). Therefore, an economic comparison between chemical consumed and benefits obtained from the anaerobic digestion for sludge without alkalis disintegration and the disintegrated sludge was presented in Table 7.

As seen in Table 7, the alkali disintegration of sludge shows a lot of benefits. Among the sludge disintegration processes, the chemical methods have advantages of easy control, stable performance and high operation flexibility. However, it is expected that the increased operation and capital costs due to chemical addition can be compensated from saving the cost of excess sludge posttreatment. In this sense, the chemical methods for sludge disintegration would be attractive and have great industrial potentials.

Table 7 An economic comparison for the alkali disintegration of sludge

Parameters	Sludge without disintegration	Disintegrated sludge	Benefits
Sludge mass reduction (%)	0 (6600 mg/L)	27% (4800 mg/L)	+
Sludge transportation and treatment (US$/ton TS)	320 (average)	234 (saved 27% due to sludge reduction)	+
Chemical consumption, g/L	0	1.2	−
Biogas enhancement, mL/g TS	245	358 (46%)	+
Electricity production, kWh (based on 1 ton TS)	524	766 (increase 46% due to biogas enhancement)	+

Note: + benefits, − negative

4 Conclusions

Sludge disintegration is used as a pretreatment step to enhance the sludge biodegradability before adding to the anaerobic digestion. The results obtained from the present study show that the sludge volume and mass were reduced significantly during digestion. MLSS reduced from 6600 mg/L down to about 4800 mg/L when alkali dosage was increased from 1.0 to 1.8 g/L. Soluble COD was increased from less than 100 mg/L to over 1800 mg/L. NaOH was an efficient reagent for inducing cell lysis and causes sludge solubilization. The sludge biodegradability enhancement is linearly correlated to COD solubilization. $Ca(OH)_2$ used for the sludge digestion could improve the sludge dewatering, but the sludge solubilization by $Ca(OH)_2$ was low. More importantly, sludge dewatering ability was increased much after sludge digestion showing that sludge management was benefited by sludge digestion. The anaerobic digestion tests show that the biogas production of the sludge without the alkali disintegration was very slow. The biogas production of the disintegrated sludge was higher compared to the sludge without the alkali disintegration (358 and 245 mL biogas/g sludge added), representing a 46% increase in the biogas production in case of NaOH disintegration. However, it was only about 10% increase in the biogas production in case of $Ca(OH)_2$ disintegration. The cost of chemical consumption for sludge disintegration was about 0.36 US $/kg TSS (about 8100 VND/kg TSS). Based on the preliminary economic assessment, if the sludge disintegration was carried out with a NaOH dose of 1.2 g/Lg, the increase in the biogas of 46% and 27% of sludge reduction could offset the cost of chemical. Therefore, the alkali sludge disintegration would be considered as a potential method for sludge management of wastewater treatment plants in Vietnam. Further works should be carried out to overcome including the reduction of chemicals and the prevention of corrosion issues.

Acknowledgments The author would like to thank the supports by the Hanoi University of Science and Technology (T2014-15). The financial supports by the GSGES seeds research funding programme for overseas field campuses (Year 2015–2016) were highly acknowledged.

References

Akbulut A (2012) Techno-economic analysis of electricity and heat generation from farm-scale biogas plant: Çiçekdağı case study. Energy 44:381–390. https://doi.org/10.1016/j.energy.2012.06.017

Amha YM, Sinha P, Lagman J, Gregori M, Smith AL (2017) Elucidating microbial community adaptation to anaerobic co-digestion of fats, oils, and grease and food waste. Water Res 123:277–289. https://doi.org/10.1016/j.watres.2017.06.065

APHA (2012) Standard methods for the examination of water and wastewater, 22nd edn. American Public Health Association/American Water Works Association/Water Environment Federation, Washington DC

Appels L, Baeyens J, Degrève J, Dewil R (2008) Principles and potential of the anaerobic digestion of waste-activated sludge. Prog Energy Combust Sci 34:755–781. https://doi.org/10.1016/j. pecs.2008.06.002

Appels L, Houtmeyers S, Degrève J, Van Impe J, Dewil R (2013) Influence of microwave pre-treatment on sludge solubilization and pilot scale semi-continuous anaerobic digestion. Bioresour Technol 128:598–603. https://doi.org/10.1016/j.biortech.2012.11.007

Aquino S, Pires EC (2016) Assessment of ozone as pretreatment to improve anaerobic digestion of vinasse. Braz J Chem Eng 33:279–285

Atay Ş, Akbal F (2016) Classification and effects of sludge disintegration technologies integrated into sludge handling units: an overview. Clean (Weinh) 44:1198–1213. https://doi.org/10.1002/clen.201400084

Baêta BEL, Lima DRS, Silva SQ, Aquino SF (2016) Influence of the applied organic load (OLR) on textile wastewater treatment using submerged anaerobic membrane bioreactors (SAMBR) in the presence of redox mediator and powered activated carbon (PAC). Braz J Chem Eng 33:817–825

Banu JR, Kavitha S (2017) Various sludge pretreatments: their impact on biogas generation. In: Singh L, Kalia VC (eds) Waste biomass management – a holistic approach. Springer International Publishing, Cham, pp 39–71. https://doi.org/10.1007/978-3-319-49595-8_3

Banu JR, Khac UD, Kumar SA, Ick-Tae Y, Kaliappan S (2012) A novel method of sludge pretreatment using the combination of alkalis. J Environ Biol 33:249–253

Bao PN, Kuyama T, Kataoka Y (2013) Urban domestic wastewater management in Vietnam – challenges and opportunities. WEPA Policy Brief 5:1–10

Barrios JA, Duran U, Cano A, Cisneros-Ortiz M, Hernández S (2017) Sludge electrooxidation as pre-treatment for anaerobic digestion. Water Sci Technol 75:775–781. https://doi.org/10.2166/wst.2016.555

Browne JD, Allen E, Murphy JD (2014) Assessing the variability in biomethane production from the organic fraction of municipal solid waste in batch and continuous operation. Appl Energy 128:307–314. https://doi.org/10.1016/j.apenergy.2014.04.097

Bruus JH, Nielsen PH, Keiding K (1992) On the stability of activated sludge flocs with implications to dewatering. Water Res 26:1597–1604. https://doi.org/10.1016/0043-1354(92)90159-2

Carrère H, Dumas C, Battimelli A, Batstone DJ, Delgenès JP, Steyer JP, Ferrer I (2010) Pretreatment methods to improve sludge anaerobic degradability: a review. J Hazard Mater 183:1–15. https://doi.org/10.1016/j.jhazmat.2010.06.129

Chen Y, Liu K, Su Y, Zheng X, Wang Q (2013) Continuous bioproduction of short-chain fatty acids from sludge enhanced by the combined use of surfactant and alkaline pH. Bioresour Technol 140:97–102. https://doi.org/10.1016/j.biortech.2013.04.075

Chiang K-Y, Chien K-L, Lu C-H (2012) Characterization and comparison of biomass produced from various sources: suggestions for selection of pretreatment technologies in biomass-to-energy. Appl Energy 100:164–171. https://doi.org/10.1016/j.apenergy.2012.06.063

Chouari R, Le Paslier D, Daegelen P, Ginestet P, Weissenbach J, Sghir A (2005) Novel predominant archaeal and bacterial groups revealed by molecular analysis of an anaerobic sludge digester. Environ Microbiol 7:1104–1115. https://doi.org/10.1111/j.1462-2920.2005.00795.x

Christodoulou A, Stamatelatou K (2016) Overview of legislation on sewage sludge management in developed countries worldwide. Water Sci Technol 73:453–462. https://doi.org/10.2166/wst.2015.521

Costa A, Gusmara C, Gardoni D, Zaninelli M, Tambone F, Sala V, Guarino M (2017) The effect of anaerobic digestion and storage on indicator microorganisms in swine and dairy manure. Environ Sci Pollut Res 24:24135. https://doi.org/10.1007/s11356-017-0011-5

Do K-U, Banu JR, Chung I-J, Yeom I-T (2009) Effect of thermochemical sludge pretreatment on sludge reduction and on performances of anoxic-aerobic membrane bioreactor treating low strength domestic wastewater. J Chem Technol Biotechnol 84:1350–1355. https://doi.org/10.1002/jctb.2189

Eriksson L, Alm B (1991) Study of flocculation mechanisms by observing effects of a complexing agent on activated sludge properties. Water Sci Technol 24:21–28

Fang W, Zhang P, Zhang G, Jin S, Li D, Zhang M, Xu X (2014) Effect of alkaline addition on anaerobic sludge digestion with combined pretreatment of alkaline and high pressure homogenization. Bioresour Technol 168:167–172. https://doi.org/10.1016/j.biortech.2014.03.050

Feki E, Khoufi S, Loukil S, Sayadi S (2015) Improvement of anaerobic digestion of waste-activated sludge by using H2O2 oxidation, electrolysis, electro-oxidation and thermo-alkaline pretreatments. Environ Sci Pollut Res 22:14717–14726. https://doi.org/10.1007/s11356-015-4677-2

Fitamo T, Treu L, Boldrin A, Sartori C, Angelidaki I, Scheutz C (2017) Microbial population dynamics in urban organic waste anaerobic co-digestion with mixed sludge during a change in feedstock composition and different hydraulic retention times. Water Res 118:261–271. https://doi.org/10.1016/j.watres.2017.04.012

Gayathri T, Kavitha S, Adish Kumar S, Kaliappan S, Yeom IT, Rajesh Banu J (2015) Effect of citric acid induced deflocculation on the ultrasonic pretreatment efficiency of dairy waste activated sludge. Ultrason Sonochem 22:333–340. https://doi.org/10.1016/j.ultsonch.2014.07.017

Grady CPL, Daigger GT, Lim HC (1999) Biological wastewater treatment, 2nd edn. Marcel Dekker, New York

Greenfield PF, Batstone DJ (2005) Anaerobic digestion: impact of future greenhouse gases mitigation policies on methane generation and usage. Water Sci Technol 52:39–47

Han X, Wang Z, Zhu C, Wu Z (2013) Effect of ultrasonic power density on extracting loosely bound and tightly bound extracellular polymeric substances. Desalination 329:35–40. https://doi.org/10.1016/j.desal.2013.09.002

Higgins MJ et al (2017) Pretreatment of a primary and secondary sludge blend at different thermal hydrolysis temperatures: impacts on anaerobic digestion, dewatering and filtrate characteristics. Water Res 122:557–569. https://doi.org/10.1016/j.watres.2017.06.016

Hirooka K, Asano R, Yokoyama A, Okazaki M, Sakamoto A, Nakai Y (2009) Reduction in excess sludge production in a dairy wastewater treatment plant via nozzle-cavitation treatment: case study of an on-farm wastewater treatment plant. Bioresour Technol 100:3161–3166. https://doi.org/10.1016/j.biortech.2009.01.011

Hosseini Koupaie E, Johnson T, Eskicioglu C (2017) Advanced anaerobic digestion of municipal sludge using a novel and energy-efficient radio frequency pretreatment system. Water Res 118:70–81. https://doi.org/10.1016/j.watres.2017.04.017

Houghton JJ, Quarmby J, Stephenson T (2001) Municipal wastewater sludge dewaterability and the presence of microbial extracellular polymer. Water Sci Technol 44:373–379

Houtmeyers S, Degrève J, Willems K, Dewil R, Appels L (2014) Comparing the influence of low power ultrasonic and microwave pre-treatments on the solubilisation and semi-continuous anaerobic digestion of waste activated sludge. Bioresour Technol 171:44–49. https://doi.org/10.1016/j.biortech.2014.08.029

Jang J-H, Ahn J-H (2013) Effect of microwave pretreatment in presence of NaOH on mesophilic anaerobic digestion of thickened waste activated sludge. Bioresour Technol 131:437–442. https://doi.org/10.1016/j.biortech.2012.09.057

Jimenez J, Gonidec E, Cacho Rivero JA, Latrille E, Vedrenne F, Steyer J-P (2014) Prediction of anaerobic biodegradability and bioaccessibility of municipal sludge by coupling sequential extractions with fluorescence spectroscopy: towards ADM1 variables characterization. Water Res 50:359–372. https://doi.org/10.1016/j.watres.2013.10.048

Jin B, Wilén B-M, Lant P (2004) Impacts of morphological, physical and chemical properties of sludge flocs on dewaterability of activated sludge. Chem Eng J 98:115–126. https://doi.org/10.1016/j.cej.2003.05.002

Kavitha S, Saranya T, Kaliappan S, Adish Kumar S, Yeom IT, Rajesh Banu J (2015a) Accelerating the sludge disintegration potential of a novel bacterial strain Planococcus jake 01 by CaCl2 induced deflocculation. Bioresour Technol 175:396–405. https://doi.org/10.1016/j.biortech.2014.10.122

Kavitha S, Yukesh Kannah R, Yeom IT, Do K-U, Banu JR (2015b) Combined thermo-chemo-sonic disintegration of waste activated sludge for biogas production. Bioresour Technol 197:383–392. https://doi.org/10.1016/j.biortech.2015.08.131

Kavitha S, Rajesh Banu J, Vinoth Kumar J, Rajkumar M (2016) Improving the biogas production performance of municipal waste activated sludge via disperser induced microwave disintegration. Bioresour Technol 217:21–27. https://doi.org/10.1016/j.biortech.2016.02.034

Kim MS, Lee K-M, Kim H-E, Lee H-J, Lee C, Lee C (2016) Disintegration of waste activated sludge by thermally-activated persulfates for enhanced dewaterability. Environ Sci Technol 50:7106–7115. https://doi.org/10.1021/acs.est.6b00019

Li C, Wang X, Zhang G, Yu G, Lin J, Wang Y (2017a) Hydrothermal and alkaline hydrothermal pretreatments plus anaerobic digestion of sewage sludge for dewatering and biogas production: bench-scale research and pilot-scale verification. Water Res 117:49–57. https://doi.org/10.1016/j.watres.2017.03.047

Li X, Xu X, Huang S, Zhou Y, Jia H (2017b) An efficient method to improve the production of methane from anaerobic digestion of waste activated sludge. Water Sci Technol 76:2075. https://doi.org/10.2166/wst.2017.313

Li Y, Hu Y, Lan W, Yan J, Chen Y, Xu M (2017c) Investigation of the accumulation of ash, heavy metals, and polycyclic aromatic hydrocarbons to assess the stability of lysis–cryptic growth sludge reduction in sequencing batch reactor. Environ Sci Pollut Res 24:24147. https://doi.org/10.1007/s11356-017-0042-y

Liu Y (2003) Chemically reduced excess sludge production in the activated sludge process. Chemosphere 50:1–7. https://doi.org/10.1016/S0045-6535(02)00551-9

Müller JA (2001) Prospects and problems of sludge pre-treatment processes. Water Sci Technol 41:121–128

Müller JA, Winter A, Strünkmann G (2004) Investigation and assessment of sludge pre-treatment processes. Water Sci Technol 49:97–104

Muz M, Ak MS, Komesli OT, Gökçay CF (2014) Intermittent ozone application in aerobic sludge digestion. Ozone Sci Eng 36:57–64. https://doi.org/10.1080/01919512.2013.824808

Nazari L et al (2017) Low-temperature thermal pre-treatment of municipal wastewater sludge: process optimization and effects on solubilization and anaerobic degradation. Water Res 113:111–123. https://doi.org/10.1016/j.watres.2016.11.055

Nga TTV (2017) Problems and challenges in nitrogen and phosphorus removal in municipal wastewater in Vietnam in Conference proceeding of "Removal of nitrogen and phosphorus in municipal wastewater in Vietnam: challenges and opportunities". Hanoi, Vietnam, pp 1–13

Nghiem LD, Koch K, Bolzonella D, Drewes JE (2017) Full scale co-digestion of wastewater sludge and food waste: bottlenecks and possibilities. Renew Sust Energ Rev 72:354–362. https://doi.org/10.1016/j.rser.2017.01.062

Nowak O (2006) Optimizing the use of sludge treatment facilities at municipal WWTPs. J Environ Sci Health A 41:1807–1817. https://doi.org/10.1080/10934520600778986

Ødegaard H, Paulsrud B, Karlsson I (2002) Wastewater sludge as a resource: sludge disposal strategies and corresponding treatment technologies aimed at sustainable handling of wastewater sludge. Water Sci Technol 46:295–303

Øegaard H (2004) Sludge minimization technologies-an overview. Water Sci Technol 49:31–40

Oh Y-K, Lee K-R, Ko K-B, Yeom I-T (2007) Effects of chemical sludge disintegration on the performances of wastewater treatment by membrane bioreactor. Water Res 41:2665–2671. https://doi.org/10.1016/j.watres.2007.02.028

Ohsaka F (2005) Activity to reduce sludge generated from septic tanks to zero using bacterial method FUJITSU. Sci Technol J 41:259–268

Park ND, Helle SS, Thring RW (2012) Combined alkaline and ultrasound pre-treatment of thickened pulp mill waste activated sludge for improved anaerobic digestion. Biomass Bioenergy 46:750–756. https://doi.org/10.1016/j.biombioe.2012.05.014

Pérez-Elvira SI, Nieto Diez P, Fdz-Polanco F (2006) Sludge minimisation technologies. Rev Environ Sci Biotechnol 5:375–398. https://doi.org/10.1007/s11157-005-5728-9

Pham TT, Mai TD, Pham TD, Hoang MT, Nguyen MK, Pham TT (2016) Industrial water mass balance as a tool for water management in industrial parks. Water Resour Ind 13:14–21. https://doi.org/10.1016/j.wri.2016.04.001

Pilli S, Yan S, Tyagi RD, Surampalli RY (2015) Thermal pretreatment of sewage sludge to enhance anaerobic digestion: a review. Crit Rev Environ Sci Technol 45:669–702. https://doi.org/10. 1080/10643389.2013.876527

Riviere D et al (2009) Towards the definition of a core of microorganisms involved in anaerobic digestion of sludge. ISME J 3:700–714. http://www.nature.com/ismej/journal/v3/n6/suppinfo/ismej20092s1.html

Romero P, Coello MD, Quiroga JM, Aragón CA (2013) Overview of sewage sludge minimisation: techniques based on cell lysis-cryptic growth. Desalin Water Treat 51:5918–5933. https://doi. org/10.1080/19443994.2013.794842

Şahinkaya S, Sevimli MF, Aygün A (2012) Improving the sludge disintegration efficiency of sonication by combining with alkalization and thermal pre-treatment methods. Water Sci Technol 65:1809–1816. https://doi.org/10.2166/wst.2012.074

Sallanko J, Okkonen J (2009) Effect of ozonation on treated municipal wastewater. J Environ Sci Health A 44:57–61. https://doi.org/10.1080/10934520802515350

Seiple TE, Coleman AM, Skaggs RL (2017) Municipal wastewater sludge as a sustainable bioresource in the United States. J Environ Manag 197:673–680. https://doi.org/10.1016/j. jenvman.2017.04.032

Song Y, Shi Z, S-y C, Luo L (2013) Feasibility of using lysozyme to reduce excess sludge in activated sludge process. J Cent South Univ 20:2472–2477. https://doi.org/10.1007/s11771-013-1759-5

Świątczak P, Cydzik-Kwiatkowska A, Rusanowska P (2017) Microbiota of anaerobic digesters in a full-scale wastewater treatment plant. Arch Environ Prot. https://doi.org/10.1515/aep-2017-0033

Tchobanoglous G, Burton FL, Stensel HD (2003) Wastewater engineering: treatment. In: Disposal and reuse, 4th edn. McGraw-Hill, New York

Thuy PT, Tuan PT, Khai NM (2016) Industrial water mass balance analysis. Int J Environ Sci Dev 7:216–220

Tyagi VK, Lo S-L (2012) Enhancement in mesophilic aerobic digestion of waste activated sludge by chemically assisted thermal pretreatment method. Bioresour Technol 119:105–113. https://doi.org/10.1016/j.biortech.2012.05.134

Uan DK, Harada H, Hoang TN, Hong NTN (2016) Application of sludge disintegration to enhance sludge management for wastewater treatment plants in Vietnam proceedings of international conference on environmental engineering and management for sustaninable development, Bach Khoa publishing house, Hanoi, 15 September, 2016. pp 23–27

Uan DK, Harada H, Saizen I (2017) Techno-economic assessment of alkalis sludge disintegration to enhance sludge management for wastewater treatment plants in Vietnam proceedings of the 7th National conference on water resources engineering, the 4th EIT international conference on water resources engineering, the 9th AUN/SEED-net regional conference on environmental engineering "development for sustainable global environment and water resources", January 23–24, 2017, Chonburi, Thailand. pp 326–330

Uma Rani R, Adish Kumar S, Kaliappan S, Yeom I-T, Rajesh Banu J (2014) Enhancing the anaerobic digestion potential of dairy waste activated sludge by two step sono-alkalization pretreatment. Ultrason Sonochem 21:1065–1074. https://doi.org/10.1016/j.ultsonch.2013.11. 007

Velho VF, Daudt GC, Martins CL, Belli Filho P, Costa RHR (2016) Reduction of excess sludge production in an activated sludge system based on lysis-cryptic growth, uncoupling metabolism and folic acid addition. Braz J Chem Eng 33:47–57. https://doi.org/10.1590/0104-6632. 20160331s20140207

Vlyssides AG, Karlis PK (2004) Thermal-alkaline solubilization of waste activated sludge as a pre-treatment stage for anaerobic digestion. Bioresour Technol 91:201–206. https://doi.org/10. 1016/S0960-8524(03)00176-7

Wang Z, Yu H, Ma J, Zheng X, Wu Z (2013) Recent advances in membrane bio-technologies for sludge reduction and treatment. Biotechnol Adv 31:1187–1199. https://doi.org/10.1016/j.biotechadv.2013.02.004

Wang R, Liu J, Lv Y, Ye X (2016a) Sewage sludge disruption through sonication to improve the co-preparation of coal–sludge slurry fuel: the effects of sonic frequency. Appl Therm Eng 99:645–651. https://doi.org/10.1016/j.applthermaleng.2016.01.098

Wang X, Duan X, Chen J, Fang K, Feng L, Yan Y, Zhou Q (2016b) Enhancing anaerobic digestion of waste activated sludge by pretreatment: effect of volatile to total solids. Environ Technol 37:1520–1529. https://doi.org/10.1080/09593330.2015.1120783

WorldBank (2013) Vietnam urban wastewater review. © World Bank, Washington, DC

WorldBank (2014) Socialist Republic of Vietnam : review of urban water and wastewater utility reform and regulation. © World Bank, Washington, DC

Yang Q, Yi J, Luo K, Jing X, Li X, Liu Y, Zeng G (2013) Improving disintegration and acidification of waste activated sludge by combined alkaline and microwave pretreatment. Process Saf Environ Prot 91:521–526. https://doi.org/10.1016/j.psep.2012.12.003

Yang Y et al (2014) Metagenomic analysis of sludge from full-scale anaerobic digesters operated in municipal wastewater treatment plants. Appl Microbiol Biotechnol 98:5709–5718. https://doi.org/10.1007/s00253-014-5648-0

Yu H, Wang Q, Wang Z, Sahinkaya E, Li Y, Ma J, Wu Z (2014) Start-up of an anaerobic dynamic membrane digester for waste activated sludge digestion: temporal variations in microbial communities. PLoS One 9:e93710. https://doi.org/10.1371/journal.pone.0093710

Yu H, Wang Z, Wu Z, Zhu C (2016) Enhanced waste activated sludge digestion using a submerged anaerobic dynamic membrane bioreactor: performance, sludge characteristics and microbial community. Sci Rep 6:20111. https://doi.org/10.1038/srep20111. https://www.nature.com/articles/srep20111#supplementary-information

Zhang G, Zhang P, Yang J, Liu H (2008) Energy-efficient sludge sonication: power and sludge characteristics. Bioresour Technol 99:9029–9031. https://doi.org/10.1016/j.biortech.2008.04.021

Zhao J et al (2017) Aged refuse enhances anaerobic digestion of waste activated sludge. Water Res 123:724–733. https://doi.org/10.1016/j.watres.2017.07.026

Zhen G, Lu X, Li Y-Y, Zhao Y (2014) Combined electrical-alkali pretreatment to increase the anaerobic hydrolysis rate of waste activated sludge during anaerobic digestion. Appl Energy 128:93–102. https://doi.org/10.1016/j.apenergy.2014.04.062

Zhen G, Lu X, Kato H, Zhao Y, Li Y-Y (2017) Overview of pretreatment strategies for enhancing sewage sludge disintegration and subsequent anaerobic digestion: current advances, full-scale application and future perspectives. Renew Sust Energ Rev 69:559–577. https://doi.org/10.1016/j.rser.2016.11.187

Development of an Iron-Based Adsorption System to Purify Biogas for Small Electricity Generation Station in Vietnam: A Case Study

Khac-Uan Do, Trung-Dung Nghiem, Shin Dong Kim, Thi-Thu-Hien Nguyen, Bich-Thuy Ly, Dac-Chi Tran, Duc-Ho Vu, and Jun Woo Park

1 Introduction

1.1 Sources and Applications of Biogas

Biogas, naturally occurring from the decomposition of all living matter, has yielded important industrial products or by-products, and its commercial value has risen for two reasons: (i) because its release into the atmosphere contributes largely to greenhouse gas concentration, with consequent and significant remediation costs, and (ii) because its energetic content is high and its exploitation means significant revenues or avoided costs (Ullah Khan et al. 2017). Systematic biogas sources linked to anthropogenic activities include nonexclusive units of landfill, commercial composting, wastewater sludge anaerobic fermentation, animal farm manure anaerobic fermentation, and agrofood industry sludge anaerobic fermentation (Surendra et al. 2014). The utilization of poultry waste for energy generation is feasible and environmentally friendly (Arshad et al. 2018). The biogas produced by all these activities is rich in CH_4 (typically ranging between 35 and 75%vol) (Abatzoglou and Boivin 2009), and its higher heating value is between 15 and 30 MJ/Nm^3 (Jiang et al. 2011; Kampanatsanyakorn et al. 2013). 1 m^3 of biogas produces hourly 2.14 kWh of electricity, and 2.47 kWh of heat energy considering the total energy value of biogas is 5.5 kWh/m^3 (Akbulut 2012). Therefore, biogas is a valuable renewable energy carrier. Biogas is becoming a promising source of renewable energy worldwide (Jin et al. 2017). It is considered as one of the main sources of energy for both developed

K.-U. Do · T.-D. Nghiem (✉) · T.-T.-H. Nguyen · B.-T. Ly · D.-C. Tran · D.-H. Vu
School of Environmental Science and Technology, Hanoi University of Science and Technology, Hanoi, Vietnam
e-mail: dung.nghiemtrung@hust.edu.vn

S. D. Kim · J. W. Park
Environment & Chemistry Solution Corporation, Seoul, Korea

© Springer International Publishing AG, part of Springer Nature 2018
H.-Y. Chan, K. Sopian (eds.), *Renewable Energy in Developing Countries*,
Green Energy and Technology, https://doi.org/10.1007/978-3-319-89809-4_10

and developing countries (Shafie et al. 2012). For example, biogas is considered an advanced biofuel in EU (European Union) and is a versatile alternative renewable form of energy, expected to contribute as vehicle fuel or for heat and electricity generation (Fitamo et al. 2017). In Sweden, biogas was produced in wastewater treatment plants (44%), at landfills (22%), and in co-digestion plants (25%). Farm-based biogas production represents approximately 1% of the total production (Lantz 2012). Biogas is one of the existing renewable energy generation capacity in Australia (Hasan Chowdhury and Maung Than Oo 2012). Biogas can be exploited directly as a fuel or as a raw material for the production of synthesis gas and/or hydrogen. Biogas produced by anaerobic digestion of manure normally consists of 50–70% methane (CH_4), 25–45% carbon dioxide (CO_2), 2–7% water (H_2O) at 20–40 °C, 2–5% nitrogen (N_2), 0–2% oxygen (O_2), and less than 1%hydrogen (H_2), 0–1% ammonia (NH_3), and 0–6000 ppm hydrogen sulfide (H_2S) (Akbulut 2012). Methane (CH_4) and carbon dioxide (CO_2) are the main constituents, but biogases also contain significant quantities of undesirable compounds (contaminants), such as H_2S, NH_3, and siloxanes (Akbulut 2012; Lu et al. 2017; Makareviciene and Sendzikiene 2015; Ullah Khan et al. 2017). The existence and quantities of these contaminants depend on the biogas source (i.e., landfills, anaerobic fermentation of manure).

Biogas from anaerobic digestion becomes an important renewable fuel that is excellent for distributed generation (Santos et al. 2016). Biogas can play a major role in the developing market for renewable energy, and it is estimated that biogas usage in the world will be doubled in the coming years ranging from 14.5 gigawatts (GW) in 2012 to 29.5 GW in 2022 (Ullah Khan et al. 2017). The energy ratios and indices for biogas production and electricity generation from the biogas showed that conversion of biogas to electricity resulted in a reduction in the relative sustainability of the system (Ciotola et al. 2011). In Mexico, the annual electricity generation is 137 GWh/a in which biogas from municipal solid waste contributes with 63%, sewage sludge with 29%, industrial organic waste with 6%, and by-products from agriculture (e.g., livestock manure) with 2% (Rios and Kaltschmitt 2016). In Pakistan, based on the waste produced in poultry forms, there is possibility to produce biogas for electricity generation. A significant addition of energy source in the existing energy system of 2.5 kWh electricity can be generated from one cubic meter of biogas (Arshad et al. 2018). Flexible power generation from biogas offers the possibility of flexible as well as controllable electricity generation and may therefore reduce electricity system transformation costs (Lauer and Thrän 2017). In case of biogas generated from waste sludge digestion, biogas and an electrical conversion efficiency of 38%, yielding an effective electricity conversion factor of 2.4–2.7 kWh/m^3 for biogas (Alvarez-Gaitan et al. 2016). In Vietnam, pig slurry is commonly applied to biogas digesters for production of biogas for electricity and cooking. The effluent from biogas digester was used to fertilize field crops, vegetables, and fish ponds (Huong et al. 2014a). Therefore, biogas potential from animal waste is huge in Vietnam. However, currently, biogas almost has been used simply for biogas burners, and lamps are designed and manufactured with high quality and acceptable costs (Pham et al. 2017).

1.2 Biogas Generation in Vietnam

Vietnam is an agricultural developing country with more than 90 million people and 12 million rural households (WorldBank 2014). In Vietnam, there is no big centralized area for breeding industry. Currently, there have been approximately 8 million of breeding households. So far, there have been about 1 million of biogas digesters in Vietnam (Pham et al. 2017). In particular, the pig production in Vietnam is rapidly increasing due to a human population and economy that are growing resulting in increased consumer demands for pork meat (Roubík et al. 2017). Pig production is projected to increase from 31.2 million pigs in 2012 to 34.8 million in 2020. In Vietnam, the amount of manure collected is also increasing. This is due not only to the increase in livestock production but also because livestock is increasingly being raised at a larger scale in animal houses (Pham et al. 2017). In consequence, Vietnam is facing huge challenges managing large volumes of pig manure in an environmentally sustainable manner (Huong et al. 2014a). Pig manure was a primary source of feedstock for small-scale biogas digesters in Vietnam (Roubík et al. 2017). Vietnam is facing many problems associated with animal waste management such as air and water pollution, lack of hygiene, and inappropriate use of manure resources. Fermentation of manure in biogas reactors is regarded as a helpful tool to solve some of these problems, and since 2003 the government has supported the construction of biogas digesters together with international organizations such as the Netherlands Development Organisation (Thien Thu et al. 2012). The current discharged waste is estimated at 75 million tons per day in total (Roubík et al. 2016, 2017). It is estimated to increase at 15% yearly.

So far, in Vietnam biogas has also been used in industrial sector as well as in small sector. For example, joint-venture Dong Nam, a beer company in Hanoi, has constructed one biogas digester with UASB style for treatment of wastewater from beer cooking technology. Wastewater volume of the factory is 600 m^3 per day. Alcohol factory of Lam Son sugarcane company in Thanh Hoa province has constructed biogas digesters with Indian technology with a volume of 16,000 m^3, treating 900 m^3 of wastewater per day. Biogas digester with Thailand UASB style with a volume of 5000 m^3 has been installed at Quang Ngai Agricultural Products and Foodstuff Joint Stock Company (Huong et al. 2014b). Besides, during urban domestic wastewater treatment, a lot of sludge generated could be used for anaerobic digestion to obtain the biogas (Bao et al. 2013; WorldBank 2013). The slaughterhouse wastewater (SHWW) in urban areas of Vietnam could be treated in anaerobic digester to produce the biogas for electricity generator. For a country like Vietnam, with a total of over 29,000 slaughterhouses divided into large (>100 m^3/day of SHWW), medium (30–100 m^3/day), and household scale (<30 m^3/day), the lack of treatment of SHWW is a major sanitary concern (Do et al. 2016). Besides, in Vietnam 35 urban WWTPs had been constructed with a total capacity of 850,000 m^3/day. Some 40 new WWTPs are in the design or construction phase with a capacity of 1,600,000 m^3/day (WorldBank 2013). During operation, a lot of excess sludge must be wasted out of the system. Therefore, the excess sludge could be combined with a low-volume flush toilet, blackwater pressure sewer, and community food residuals to add into the anaerobic digester as a co-digestion process. The disposal of sewage sludge and food waste could be the best option regarding the reduction of greenhouse gas emissions (Lijó et al. 2017). Co-digestion and mixing, e.g., sewage sludge with food waste, give a higher methane yield than if these two were digested separately; the increase could be as high as 60%, depending on the substrate used (Cobbledick et al. 2016; Hasan et al. 2012; Kim et al. 2016; Mohseni et al. 2012; Nghiem et al. 2017). The benefit from energy generated by the anaerobic digester could be achieved (Schoen et al. 2017). However, only about 30–50% of the volatile solids containing in sludge can be degraded through anaerobic digestion, indicating a low overall energy efficiency of anaerobic digestion (Gu et al. 2017). Therefore, it could be seen that the biogas has been developed strongly in Vietnam. This leads to a good chance to use biogas and to develop the commerce of biogas technology. One of the important steps is to purify the biogas to improve the quality of biogas before using such as for cooking, lighting, and specially as fuel for running internal combustion engines or power generator. Currently, in Vietnam market, the users can buy the filters to install into biogas systems to purify the biogas. However, biogas user's awareness of biogas purification is still limited. At household scale $(10–15 \text{ m}^3)$, there are about 1,000,000 biogas digesters in which each digester can produce about 1000 m^3/year. Besides, there are about 17,000 pig farms with over 500 pigs, and less than 0.3% of them have a biogas facility. Biogas production from this size is about 50,000 m^3/year. Therefore, this will make a large market for biogas purification for small electricity generation station commercialization.

1.3 Current Status and Challenges of Biogas Digesters in Vietnam

It can be seen that there is a great potential of biogas development in Vietnam, and as with over 17,000 existing farms and fast livestock growth production, biogas potential in the sector is very high (Thien Thu et al. 2012). Besides, the wastes from sugar industry, cassava processing factories, fruit, export canned food, beer and refreshment industries, and leachate from the domestic and urban solid waste landfills could be provided for the anaerobic digesters for biogas production (Bao et al. 2013; Roubík et al. 2017). However, there are also a number of challenges for the development of biogas in Vietnam:

(i) Biogas digester is still in small scale, and there is still lack of large-scale biogas since the technologies for it require high cost.

(ii) Biogas digesters can be very instrumental for waste treatment; however if the operation of the system is not correctly done, it could cause damage or shorten the life of the biogas digester as well as its auxiliary systems. Gas production is an important indicator to evaluate the operation process of biogas digesters. It is

Table 1 A classification of biogas digester in Vietnam

No.	Size of digester	Volume of digester (V), m^3	References
1	Large (industrial scale)	$V > 1000$	Cheng et al. (2014)
2	Medium (farm scale)	$50 < V < 1000$	Rennuit and Sommer (2013)
3	Small (family scale)	$V < 50$	Bruun et al. (2014)

not only affected by the operation process but also the cold weather conditions in the north of Vietnam.

(iii) Technical barriers in operation and maintenance of biogas digester, especially on gas purification.

(iv) Market for biogas has not been developed.

It should be noted that biogas production in large scale and commercial system (i. e., from agro-agricultural sectors, sugarcane, cassava starch, rubber) is necessary for Vietnam in the coming years. However, it was not widely applied, since most of biogas are small sized and the amount of gas produced is insufficient for gas generator. It should be noted that biogas production in large scale and commercial system (i.e., from agro-agricultural sectors, sugarcane, cassava starch, rubber) is necessary for Vietnam in the coming years. Medium-scale biogas digesters have been applied widely for farm scale. The biogas digester could be seen as a small scale if its volume was smaller than 40 m^3, or if the size of the piggery breeding was less than 100 pigs (Ghimire 2013; Thien Thu et al. 2012). A classification of biogas digester size in Vietnam is presented in Table 1.

In small sector, biogas has been also developed based on the simple material, such as plastic biogas digesters or brick masonry compartment (Roubík et al. 2016). Most small-scale biogas digesters are constructed below ground (Bruun et al. 2014). Simple design, ease of installation, and low specialized manpower demands make polyethylene tubular digester both affordable and acceptable for household applications in developing countries such as Colombia, Ethiopia, Tanzania, Vietnam, Cambodia, Costa Rica, Bolivia, Peru, Ecuador, Argentina, Chile, and Mexico where installation of the fixed dome and floating drum digesters has been reported to be too costly (Surendra et al. 2014). Actually, the anaerobic digestion of animal manure has been practiced in Vietnam since the 1960s (Huong et al. 2014b). Vietnam has tested low-cost polyethylene tube digesters since the 1980s. Currently, about two million households have already installed biogas digesters in Vietnam, including at least one million low-cost polyethylene plastic biogas digesters. In Vietnam, low-cost polyethylene tube biodigester has been proven to be a cheap and simple method of producing biogas for small-scale farms (Cheng et al. 2014). It should be noted that in Vietnam, 25–32% of biogas digesters have hydraulic retention time shorter than 10 days, which is not optimal for biogas production (Pham et al. 2014). Biogas technology in Vietnam remains primarily focused on family-sized biogas technology (Roubík et al. 2017). Energy consumption in the rural areas of Central Vietnam can be met by the use of household-sized biogas digesters and can provide a healthier and more sustainable way of living. Biogas was collected and used for several purposes, but it is mainly used for cooking or lighting (Bruun et al. 2014; Ghimire 2013; Thien Thu et al. 2012). However, it was reported from a survey of four different provinces (Binh Dinh, Dong Nai, Gia Lai, and Tra Vinh) in Southern Vietnam that 140 of 216 households (65%) with a biogas digester had excess biogas that they could not use in summer. Of these 140 farms with excess biogas, 68 (48.6%) released it directly into the atmosphere, while the rest either gave or sold it to a neighbor or burned it (Bruun et al. 2014). The smell of biogas was mentioned as another problem and was identified in households without a desulfurization unit or that did not perform proper maintenance of that unit (Roubík et al. 2016).

1.4 The Needs for Biogas Purification in Electricity Generation

As mentioned above, biogas contains significant quantities of undesirable compound, such as hydrogen sulfide (H_2S) (Abatzoglou and Boivin 2009). H_2S is corrosive, toxic, and odorous. So it can significantly damage mechanical and electrical equipment used for process control, energy generation, and heat recovery (Ciotola et al. 2011). The combustion of H_2S results in the release of sulfur dioxide, which is a problematic environmental gas emission (Dornelas et al. 2017). Boilers, which generate heat from biogas, do not have a high gas quality requirement, although it is recommended that H_2S concentrations be kept below 1000 ppm. Internal combustion engines, used for electricity generation, have required biogas with H_2S concentrations below 100 ppm (Arshad et al. 2018; Dornelas et al. 2017). Biogas can also be utilized as a vehicle fuel. However, it must be upgraded because vehicles need a much higher gas quality. H_2S can be controlled using a variety of methods which can be either physical-chemical or biological removal processes (Jiang et al. 2011). Activated carbons are frequently used for biogas adsorption because of their high surface area, porosity, and surface chemistry where H_2S can be physically and chemically adsorbed. Much of the research has focused on how the physical and chemical properties of various activated carbons affect the breakthrough capacity of H_2S. Activated carbon can be impregnated with potassium hydroxide (KOH) or sodium hydroxide (NaOH), which acts as catalyst to remove H_2S (Álvarez-Gutiérrez et al. 2014; Kampanatsanyakorn et al. 2013). Much research has focused on mechanisms of H_2S removal using activated carbon. The studies have focused on hydrogen sulfide adsorption on activated carbons as it relates to surface properties, surface chemistry, temperature, concentration of H_2S gas, addition of cations, moisture of gas stream, and pH. These experiments have used both biogas from real processes and laboratory-produced gases of controlled composition (Matassa et al. 2015; Surendra et al. 2014). NaOH-impregnated activated carbon was also tested for H_2S removal capacity. The results showed that with increasing amounts of NaOH added, the H_2S removal capacity of the activated carbons increases (Abatzoglou and Boivin 2009). Zeolites are especially effective at removing polar compounds, such as water and H_2S, from nonpolar gas streams, such as methane (Ciotola et al. 2011). The use of zeolite NaX and zeolite KX as catalysts for removing H_2S from biogas streams has been performed. The study found a yield of 86% of elemental sulfur on the zeolites over a period of 40 h (Jiang et al. 2011). When sludge undergoes pyrolysis, a material is obtained with a mesoporous structure and an active surface area with chemistry that may promote the oxidation of H_2S to elemental sulfur. Samples with higher content of sewage sludge pyrolyzed at higher temperatures (800 and 950 °C) had the best adsorption capacity (Ullah Khan et al. 2017). Metal oxides have been tested for hydrogen sulfide adsorption capacities. Iron oxide is often used for H_2S removal. It can remove H_2S by forming insoluble iron sulfides. Iron oxide can be used in either a batch system or a continuous system. In a continuous system, air is continuously added to the gas stream so that the iron oxide is regenerated simultaneously. In a batch mode operation, where the iron oxide is used until it is completely spent and then replaced, it has been found that the theoretical efficiency is approximately 85% (Abatzoglou and Boivin 2009).

In Vietnam, there is still a lack of information of biogas purification technology. This will make a large chance for study and apply new technology for biogas purification, such as removal of H_2S and CO_2 from biogas. Another issue that hinders the application of biogas development in Vietnam is that there is no information on siloxane removal from biogas. Siloxanes are widely employed by industry because of their interesting properties, including low flammability, low surface tension, thermal stability, hydrophobicity, high compressibility, and low toxicity. Moreover, they are not environmentally persistent compounds and generally have very low allergenicity. Among others, they can be found in shampoos, pressurized cans (i.e., hair sprays, shaving foams), detergents, cosmetics, pharmaceuticals, textiles, and paper coatings. During the anaerobic digestion of waste sludges and in landfills, siloxanes do not decompose; they are significantly volatilized and, thus, they are transferred to biogas. The main problem with siloxanes in biogas is that they produce microcrystalline silica (MCS) when biogas is used as an energy vector (during combustion). MCS has glass properties, and the fouling of metallic surfaces leads to abrasion, ill-functioning spark plugs, overheating of sensitive parts of engines due to coating, and general deterioration of all mechanical engine parts (Abatzoglou and Boivin 2009). More importantly, the presence of high H_2S constitutes a major problem because (i) they can be detrimental to any biogas thermal or thermocatalytic conversion device (e.g., corrosion, erosion, fouling) and (ii) they generate harmful environmental emissions (Ács et al. 2015; Costa et al. 2017; Crone et al. 2016; Dornelas et al. 2017; Jensen et al. 2014; Kavitha et al. 2015; Kovács et al. 2015; Lauer and Thrän 2017). Besides, it has been reported that free H_2S has a critical effect on the anaerobic treatment of methanol-containing wastewater. Due to its ability to penetrate cells and block metalloproteins, high sulfide, for example, reduces the efficiency of biomethanation and even destroys bacteria, including sulfate-reducing bacteria (Lu et al. 2017). So, it is important to include biogas purification steps upstream of its final use processes.

1.5   Objectives and Scopes of This Work

It should be noted that in Vietnam, biogas has been used for electricity generation, but still at small scale of about 0.5–20 kW. The use of biogas for electricity generation at household scale remains very low, with only 500 out of a total of 200,000 biogas facilities installed. It should be noted that a gasoline generator modified to run on biogas or as a hybrid unit consumes 0.6–0.7 m^3 of biogas for 1 kWh of power generated. The capacity need for a farm is often 8–20 kW and capable of running from 6 to 10 h per day. Therefore, 30–200 m^3 of biogas will be suitable for power generation. As the electrical power generated by biogas-based machines is alternating currents, it could be connected directly to the distribution system of a regular power grid, or it could be used in an independent distribution system.

Therefore, in this study, biogas produced by the anaerobic digesters at the Thanh Hung pig farm (Thanh Oai District, Hanoi) was used for (i) testing in a small-scale adsorption system which was conducted to operate for long term to confirm the biogas purification efficiency by using the iron-based (FeOOH) adsorbent; (ii) then a pilot plant was developed to produce clean biogas for electricity generator. The electricity generated was used for many purposes at the farm.

2   Materials and Methods

2.1   Research Location

In this study, biogas generated from Thanh Hung pig farm (located in Thanh Oai District, a suburb of Hanoi) was used. This pig farm is far from Hanoi center of about 25 km. Figure 1 shows a map of the Thanh Hung pig farm. This pig farm has about 4500 pigs per year. The average wastewater was about 300 m^3/day, and the solid wastes were 1500–2000 kg/day. All of wastewater containing solid wastes was collected and introduced into a covered anaerobic lagoon with its dimensions as $L \times W \times H = 36.5 \times 14.5 \times 4.0$ m. In this, the biogas storage part including the height of about 2.5 m, corresponding to biogas storage volume, varied from 490 to over 800 m^3, depending on the amount of wastewater and amount of biogas used. It should be noted that biogas without purification was mostly used for cooking at the farm (Fig. 2).

2.2   Lab-Scale H₂S Purification System

A lab-scale system was set up at the Thanh Hung pig farm (Thanh Oai District, Hanoi). The schematic of the adsorption system is shown in Fig. 3. The actual setup

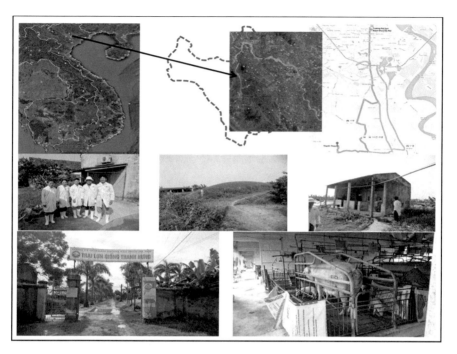

Fig. 1 Location of research area in Hanoi

Fig. 2 Biogas was used for cooking at Thanh Hung pig farm

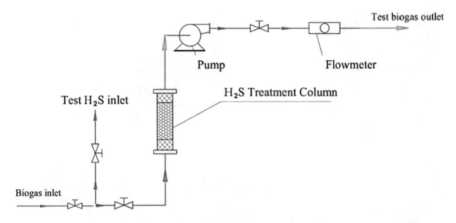

Fig. 3 The experimental system for biogas purification

Fig. 4 Pictures of the lab-scale system at Thanh Hung pig farm

of the experimental system is shown in Fig. 4. The biogas from the anaerobic digesters at the Thanh Hung pig farm entered the system. The biogas was then either sampled to measure the inlet H_2S concentration if the inlet sampling valve opened or continued to pass through a rotameter and into the bottom of the adsorbent column (inside diameter of 45 mm, height of 90 mm). The rotameter was used to control the flow rate of biogas through the system. Once the biogas entered the adsorber column, it passed through the media bed. There was also a gas sample port located in the outlet of the adsorption column for measuring the purified biogas compositions. In this study, the iron-based adsorbent (FeOOH, called as hydrated goethite or iron oxyhydroxide) developed and provided by E&Chem Solution Co. (Korea) was used for testing. It has a particle size ranging from 0.5 mm to 1.18 mm with the bulk density of 0.67 g/mL. The amount of adsorbent used was 64.35 g for testing the effect of adsorbent mass on H_2S removal. In this system, the biogas flow rate was maintained at 6 L/h.

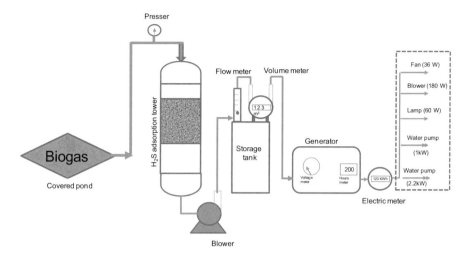

Fig. 5 Pilot plant of H₂S adsorption system and electricity generator in this study

2.3   Pilot Plant H₂S Purification System

Based on the results obtained from the small system, a bigger H_2S purification tower $(D \times H = 700 \times 1400)$ was manufactured and installed at the Thanh Hung pig farm (Fig. 5).

2.4   Electricity Generator

In this study, electricity generator CC5000-MG/LPG-T2 was used (Fig. 6). The power capacity is 3.5 kW. Main biogas valve should be equipped on main biogas pipe. Recommended gas pressure was from 2.0 to 6.0 kPa with flow rate higher than 1 m³/h. The generator engine was started by the following steps.

- Step 01 Open gas supply source.
- Step 02 Turn ENGINE SWITCH at ON position.
- Step 03 For recoil start, pull the STARTER GRIP slowly until resistance is felt and then pull rapidly. Or for electric start, turn the key to START position and hold until engine started. If the engine fails to start within 5 s, release the key and wait at least 5 s before attempting to start the engine again.
- Step 04 After the engine started, return the STARTER GRIP gently to prevent damage to the starter or housing.
- Step 05 Turn the CHOKE VALVE GRIP at OPEN or RUN position. When engine warms up, move the CHOKE VALVE GRIP to RUN or OPEN position.
- Step 06 Connect load to the generator set.
- Step 07 Turn CIRCUIT BREAKER at ON position.

Fig. 6 Start the generator engine

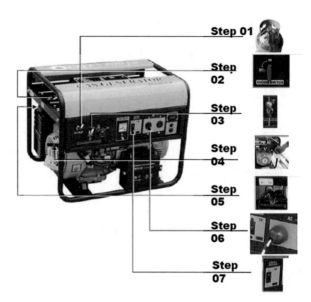

2.5   Analytical Methods

A rotameter (Yinhuan Co., Taiwan) was used to control the biogas flow rate. It had the capability of measuring flows between 0.5 and 5 L/min. Besides, a biogas volume meter (Model G1.6, Shinhan Co., Korea) was used to record the total biogas which was purified. Temperatures and humidities in the inlet and outlet were measured by the digital temperature/humidity meters (Wellink HL101, Taiwan). Samples of the inlet and outlet (purified) biogas were taken during experimental tests. The biogas composition was taken during experimental tests. The compositions of these samples (CH_4, CO_2, H_2S, O_2) were determined using an Optima 7 Biogas Analyzer (MRU Instruments, Inc., Houston, Texas, USA). The biogas compositions were also periodically measured and analyzed in the laboratory following APHA (APHA 2012). Besides, siloxanes in biogas have been collected into the Tedlar bags and to be analyzed by using GC/MS at the laboratory. Tables 2 and 3 show the siloxane analysis method and main compositions of siloxanes. In addition, during operation, the electricity generated was measured and recorded by using a kWhr meter (EMIC Corp., Vietnam).

Table 2 Siloxane analysis method

Items	Conditions	
GC model	GC/MS (HP7820A/HP5977E)	
Injector	Injection method	Manual
	Injection volume	300 μL
	Temperature	250 °C
	Split ratio	1:20
	Liner	Split focus liner (SGE, AUS)
Carrier gas	He	1 mL/min
Column	HP-5, 30 m × 0.25 mmI.D × 0.25 μm	
Detector	MS source	230 °C
	MS quad	150 °C
Oven program	Initial temperature	40 °C at 4 min
	Ramp rate	10 °C/min
	Final temperature	250 °C at 0 min

Table 3 Group of siloxanes

Ion group	Substance*	M.W.	R.T.	m/z
1	L2	162	2.8	147
2	L3	236	7.48	221
3	D4	296	9.86	281
4	L4	310	11.12	207, 295
5	D5	370	12.53	267, 355
6	L5	384	13.88	147, 281, 369

*Note: L2, MM: hexamethyldisiloxane
L3, MDM: octamethyltrisiloxane
L4, MD2M: decamethyltetrasiloxane
L5, MD3M: dodecamethylpentasiloxane
D4: octamethylcyclotetrasiloxane
D5: decamethylcyclopentasiloxane

3 Results and Discussion

3.1 Characteristics of Biogas in This Study

Thanh Hung pig farm (at Thanh Oai District, Hanoi) has about 4500 pigs per year. The average wastewater was about 300 m^3/day, and the solid waste was 1500–2000 kg/day. The wastewater and solid wastes were collected and introduced into a covered anaerobic lagoon with its dimensions of $L \times W \times H = 36.5 \times 14.5 \times 4.0$ m, in which the biogas storage part including the height of about 2.5 m, corresponding to biogas storage volume, varied from 490 to over 800 m^3, depending on the amount of wastewater and amount of biogas used. The initial biogas characteristics were analyzed. As a result, CH$_4$ was about 72%, showing that the biogas at Thanh Hung pig farm has high CH$_4$ concentration and it could be good for usage. However, the H$_2$S was too high from 1995 to over 4000 ppm. The analyzed result of

Table 4 The analyzed results of siloxanes

Sample name	Sampling method	L2	L3	L4	L5	D4 (mg/Nm^3)	D5
A1	Tedlar bag	N.D.	N.D.	N.D.	N.D.	N.D.	N.D.
A2	Tedlar bag	N.D.	N.D.	N.D.	N.D.	N.D.	N.D.
B	Tedlar bag	N.D.	N.D.	N.D.	N.D.	N.D.	N.D.
C1	Impinger absorption	N.D.	N.D.	N.D.	N.D.	1.66	N.D.
C2	Impinger absorption	N.D.	N.D.	N.D.	N.D.	1.54	N.D.

Note: *N.D.* not detected

siloxanes was shown in Table 4. It is interesting that the linear siloxane (L1–L5) concentrations in the biogas generated from Thanh Hung pig farm were almost not detected. However, the cyclic siloxanes (cyclomethicones) (D4) were detected at the level of 1.54–1.66 mg/Nm^3. The main problem with siloxanes in biogas is that they produce microcrystalline silica (MCS) when biogas is used during combustion. MCS has glass properties, and the fouling of metallic surfaces leads to abrasion, ill-functioning spark plugs, overheating of sensitive parts of engines due to coating, and general deterioration of all mechanical engine parts (Abatzoglou and Boivin 2009). Due to the low quality of biogas, currently, in Vietnam the biogas was mostly used for cooking. Sometimes, it was discharged directly to the atmosphere.

3.2 *H$_2$S Purification at Lab-Scale System*

3.2.1 Flow Rate and Cumulated Volume

The flow rate of biogas during operation of the system was recorded and presented in Fig. 7. Initially, the flow rate was controlled at about 1 L/min. After 374 min of operation, over 350 L of biogas was purified. During operation, the variation of temperature and moisture in biogas before and after purification was measured and presented in Fig. 8. The temperature in the air and in biogas was variably stable in the range of 29–32 °C. However, the moisture was much different between in the air and in biogas. During the time of experiment, the moisture in the air was around 56%, whereas it was about 87% in biogas. The high moisture will affect the H$_2$S adsorption.

Figure 9 shows the change of adsorbent during operation. It was clear that after about 374 min of operation, the adsorbent was changed from brown to black. However, it was only for the adsorbent outside. The center of adsorbent still adsorbs more H$_2$S.

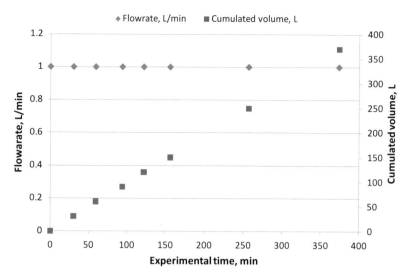

Fig. 7 Flow rate and cumulative volume of biogas during operation

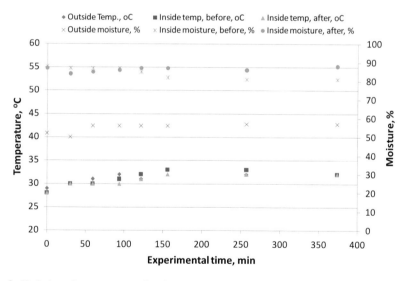

Fig. 8 Variation of temperature and moisture in biogas before and after purification

3.2.2 Variation of H₂S During Purification

After 374 min of operation, the H_2S concentration was almost lower than 0.5 ppm. After 250 min of operation, it started to increase. After 370 min, it increased to about 3 ppm. It should be noted that H_2S in biogas electric generator engine should be less than 200 ppm. It could be estimated based on adsorbent capacity that when 17.5 g

Fig. 9 The change of adsorbent during operation

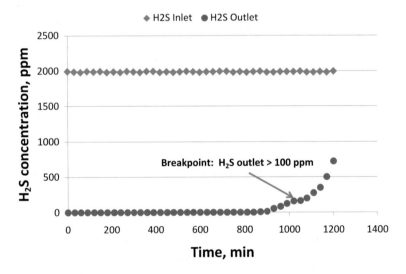

Fig. 10 Variation of H_2S during purification

FeOOH was used, it could adsorb 3.2 g H_2S; therefore the system could last the operation time of up to 1068 min (Fig. 10).

After adsorption, the adsorbent was taken out of the column and put in the air. After about 1 h, most of adsorbent was broken down easily (Fig. 11).

Normally, biogas is fully saturated with water vapor. It should be noted that the water vapor was condensed at the lower temperature. When the gas warms up again, its relative vapor content decreases. For family digester, water trap can be used to

Fig. 11 Adsorbent after H_2S purification and contact with air for about 1 h

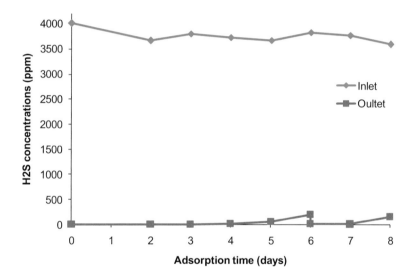

Fig. 12 Variation of H_2S inlet and outlet

remove water vapor. It is simply using natural method by installing gas pipeline with lower part in digester side and higher in biogas appliance side.

The adsorption velocity was maintained at 6 L/h during the first period from April 5 to April 14, 2016. The adsorption velocity was maintained at 6 L/h. Before adsorption, the FeOOH mass was used at 64.35 g. After adsorption, Before adsorption, the FeOOH mass was used at 64.35 g. The variation of H_2S inlet and outlet was presented in Fig. 12.

It could be seen from Fig. 13 that the H_2S concentration in the inlet was very high. It was in the range of 3500 to over 4000 ppm. The H_2S removal was effective. The H_2S concentration in the outlet was increased slowly in the first 4 days, from 0 to

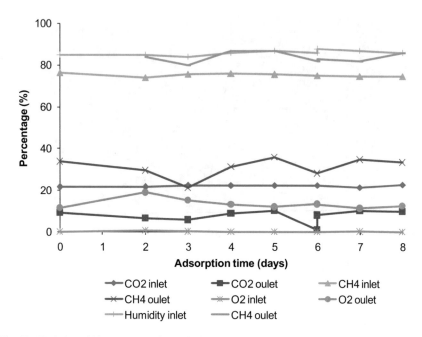

Fig. 13 Variation of biogas compositions during operation

55 ppm. However, it was still two times lower compared to the limit H_2S concentration in biogas for electrical generation (100 ppm). However, O_2 was detected in the outlet due to the adsorption system being unclosed. The reason could be due to a leaking from a blower. Thus, the experiment has been stopped after 8 days of operation.

The adsorption velocity was also maintained at 6 L/h by a vacuum pump during the second period (from April 29 to May 30, 2016), a similar condition was used. The adsorption velocity was maintained at 6 L/h by a vacuum pump. The FeOOH weight before adsorption was 62.55 g. The FeOOH weight after adsorption was increased to 99.06 g. In this period, the blower was replaced by a vacuum pump. As seen in Fig. 14, after replacing blower with vacuum pumper, the adsorption system was closed completely, and the O_2 was not detected. Figure 15 presents the variation of H_2S in the inlet and outlet. In this period, the H_2S in the inlet was lower than the first period. It fluctuated from 2100 to 3500 ppm. During operation of 30 days, the H_2S removal was very effective. The H_2S in the outlet was lower than 100 ppm after more than 22 days of operation continuously. Then slowly it was increased up to over 500 ppm after 30 days.

The flow rate of biogas during operation of the system was recorded. The flow rate was controlled at about 6 L/h. After 30 days of operation, over 4.3 m^3 of biogas was purified. It should be noted that, during operation, the temperature in the air and in biogas was variably stable in the range of 29–32 °C. Figure 16 shows the change of adsorbent during operation. It was clear that after about 30 days of operation, the

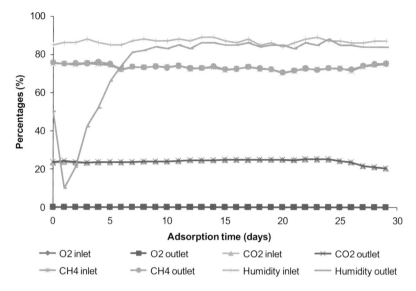

Fig. 14 Variation of biogas compositions during operation

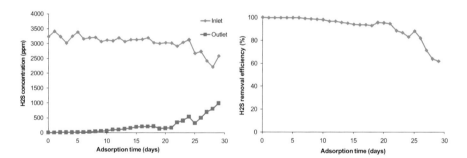

Fig. 15 Variation of H_2S inlet/outlet and H_2S removal efficiency

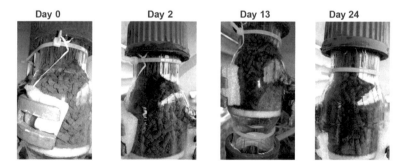

Fig. 16 The change of adsorbent during operation

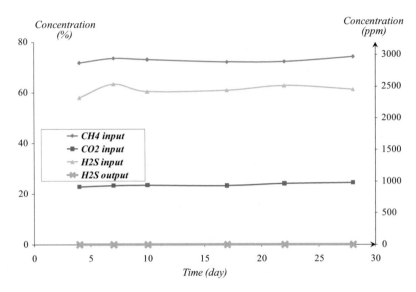

Fig. 17 Variation of biogas compositions during operation.

adsorbent was changed from brown to black. However, it was only for the adsorbent outside. The inside (center) of adsorbent was still able to adsorb more H_2S.

3.3   Pilot-Scale System for H_2S Purification and Electricity Generation

3.3.1   Variation of Biogas Compositions

During operation, the biogas compositions were frequently analyzed. As shown in Fig. 17, after 30 days of operation, the H_2S in the inlet was varied around 3000 ppm, and the H_2S in the outlet was almost zero. The results confirmed that the H_2S was removed completely by the adsorbent FeOOH produced and supplied by E&Chem Solution Company (Korea).

3.3.2   Variation of Operation Time and Biogas Used per Day

During operation, an hour meter records the total running hours the generator set has been used. With the help of hour meter, we can easily follow the scheduled maintenance activities and assure longer life of your generator set. As recommended by the manufacturer, the operation time should be around 8 h per day. The actual operation time and variation of biogas used and electric produced were presented in Fig. 18. From Fig. 18, it could be seen that the operational time was varied from 2.5

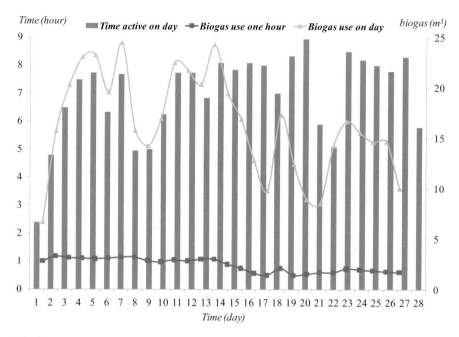

Fig. 18 Variation of operation time and biogas used per day

to about 9 h per day. It was depending on the available time of worker and staff. It was also depending on the electricity user's purposes. The average biogas flow rate used was about 1.5–3.2 m³/h, and the biogas volume used was about 10 to 25 m³/day.

3.3.3 Variation of Biogas Used and Electricity Generation

Figure 19 shows that depending on the operational time, the generator could produce from 20 to more than 30 kW per day. However, the actual electricity used was only about 5 to more than 10 kW per day. This is because, during that time, the farm only used electricity for light, fan, and water pump. It means the farm can increase the usage purposes.

Figure 20 presents the overall biogas used and electric generated for about 30 days. It is clear that the total biogas used was about 450 m³ after more than 200 h of operation. Total electric generated was more than 700 kW. It means 0.64 m³ biogas was used to generate 1 kW of electricity. Actually, there were only about 200 kW used at the farm among the 700 kW produced. It should be noted that in Vietnam, the use of biogas for electricity generation at household scale remains very low, with only 500 out of a total of 200,000 biogas facilities installed. The capacity need for a farm is often 8–20 kW and capable of running from 6 to 10 h per day. Therefore, 30–200 m³ of biogas will be suitable for power generation.

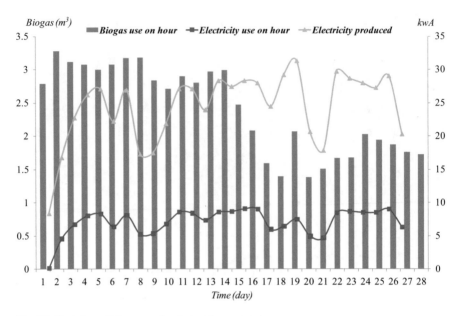

Fig. 19 Variation of biogas used and electric generated

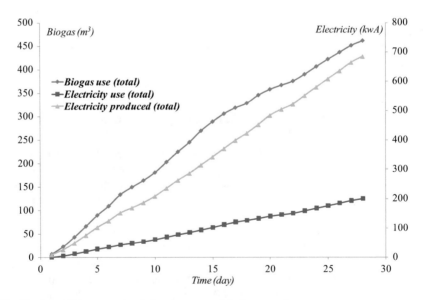

Fig. 20 Overall biogas used and electric generated during operation

4 Conclusions

Based on the results obtained from the study, it could be concluded that FeOOH has high capacity for H_2S purification from biogas. The biogas after purification using FeOOH was suitable for electric generator. During operation, the estimation of the consumed biogas and electricity produced was about 0.64 m^3 of biogas used per 1 kWh produced. The biogas flow rate varied from 2 to 2.5 m^3/h which means 48–60 m^3/day used can produce 84 kWh per day. This amount of electricity production could be fully used for a farm activity. Therefore, biogas purification technology plays an important role in biogas market in Vietnam. In Vietnam, there are over 17,000 existing farms with fast livestock growth production; biogas potential in the sector is very high. Besides, industries in Vietnam which have high organic waste including sugar industry, cassava processing factories, fruit, export canned food, beer and refreshment industries, and domestic and urban solid waste landfills are suitable industries to apply anaerobic treatment processing and biogas production. The potential market for biogas production and electricity generator from biogas in Vietnam is huge.

Acknowledgments This work was sponsored by the Korean government's International Cooperation Program for Environmental Technologies. The fund from Korea Environmental Industry and Technology Institute (KEITI) was acknowledged. The authors also would like to thank the help of people in Thanh Hung pig farm.

Supporting Information

Picture 1 Anaerobic cover lagoon at Thanh Hung pig farm

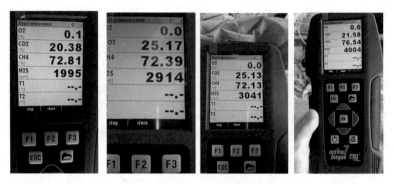

Picture 2 Analysis results of biogas compositions from Thanh Hung pig farm

Picture 3 Iron-based
adsorbents (FeOOH) used in
this study

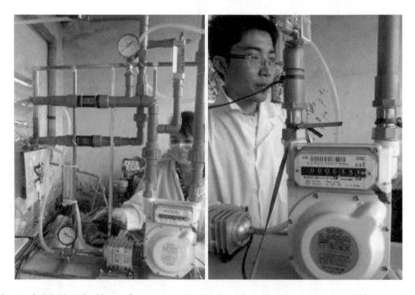

Picture 4 Monitor the biogas flow rate, temperature, moisture, pressure gauge, and biogas volume

Picture 5 Biogas before and after purification was sampled

Picture 6 Adsorbent tower was designed and manufactured

Picture 7 H$_2$S adsorption system installed at the Thanh Hung pig farm

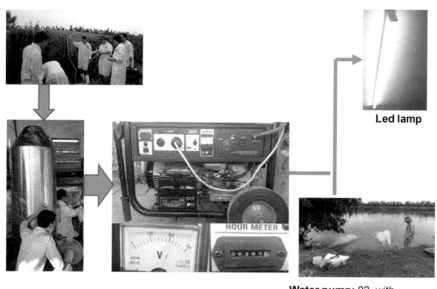

Led lamp

Water pump: 02, with capacity: 1 and 2.2 kW, Vacuum blower (180 W), Fan (36 W)

Picture 8 H$_2$S adsorption system and electricity generator installed at the Thanh Hung pig farm

References

Abatzoglou N, Boivin S (2009) A review of biogas purification processes. Biofuels Bioprod Biorefin 3:42–71. https://doi.org/10.1002/bbb.117

Ács N, Bagi Z, Rákhely G, Minárovics J, Nagy K, Kovács KL (2015) Bioaugmentation of biogas production by a hydrogen-producing bacterium. Bioresour Technol 186:286–293. https://doi.org/10.1016/j.biortech.2015.02.098

Akbulut A (2012) Techno-economic analysis of electricity and heat generation from farm-scale biogas plant: Çiçekdağı case study. Energy 44:381–390. https://doi.org/10.1016/j.energy.2012.06.017

Alvarez-Gaitan JP, Short MD, Lundie S, Stuetz R (2016) Towards a comprehensive greenhouse gas emissions inventory for biosolids. Water Res 96:299–307. https://doi.org/10.1016/j.watres.2016.03.059

Álvarez-Gutiérrez N, García S, Gil MV, Rubiera F, Pevida C (2014) Towards bio-upgrading of biogas: biomass waste-based adsorbents. Energy Procedia 63:6527–6533. https://doi.org/10.1016/j.egypro.2014.11.688

APHA (2012) Standard methods for the examination of water and wastewater, 22nd edn. American Public Health Association/American Water Works Association/Water Environment Federation, Washington DC

Arshad M, Bano I, Khan N, Shahzad MI, Younus M, Abbas M, Iqbal M (2018) Electricity generation from biogas of poultry waste: an assessment of potential and feasibility in Pakistan. Renew Sust Energ Rev 81:1241–1246. https://doi.org/10.1016/j.rser.2017.09.007

Bao PN, Kuyama T, Kataoka Y (2013) Urban domestic wastewater management in Vietnam – challenges and oppotunities. WEPA Policy Brief 5:1–10

Bruun S, Jensen LS, Khanh Vu VT, Sommer S (2014) Small-scale household biogas digesters: an option for global warming mitigation or a potential climate bomb? Renew Sust Energ Rev 33:736–741. https://doi.org/10.1016/j.rser.2014.02.033

Cheng S, Li Z, Mang H-P, Huba E-M, Gao R, Wang X (2014) Development and application of prefabricated biogas digesters in developing countries. Renew Sust Energ Rev 34:387–400. https://doi.org/10.1016/j.rser.2014.03.035

Ciotola RJ, Lansing S, Martin JF (2011) Emergy analysis of biogas production and electricity generation from small-scale agricultural digesters. Ecol Eng 37:1681–1691. https://doi.org/10.1016/j.ecoleng.2011.06.031

Cobbledick J, Aubry N, Zhang V, Rollings-Scattergood S, Latulippe DR (2016) Lab-scale demonstration of recuperative thickening technology for enhanced biogas production and dewaterability in anaerobic digestion processes. Water Res 95:39–47. https://doi.org/10.1016/j.watres.2016.02.051

Costa A, Gusmara C, Gardoni D, Zaninelli M, Tambone F, Sala V, Guarino M (2017) The effect of anaerobic digestion and storage on indicator microorganisms in swine and dairy manure. Environ Sci Pollut Res 24:24135. https://doi.org/10.1007/s11356-017-0011-5

Crone BC, Garland JL, Sorial GA, Vane LM (2016) Significance of dissolved methane in effluents of anaerobically treated low strength wastewater and potential for recovery as an energy product: a review. Water Res 104:520–531. https://doi.org/10.1016/j.watres.2016.08.019

Do AT, Bach DQ, Do UK, Prieto A, Lan Huong HT (2016) Performance of airlift MBR for on-site treatment of slaughterhouse wastewater in urban areas of Vietnam. Water Sci Technol 74:2245. https://doi.org/10.2166/wst.2016.418

Dornelas KC, Schneider RM, do Amaral AG (2017) Biogas from poultry waste—production and energy potential. Environ Monit Assess 189:407. https://doi.org/10.1007/s10661-017-6054-8

Fitamo T, Triolo JM, Boldrin A, Scheutz C (2017) Rapid biochemical methane potential prediction of urban organic waste with near-infrared reflectance spectroscopy. Water Res 119:242–251. https://doi.org/10.1016/j.watres.2017.04.051

Ghimire PC (2013) SNV supported domestic biogas programmes in Asia and Africa. Renew Energy 49:90–94. https://doi.org/10.1016/j.renene.2012.01.058

Gu J, Xu G, Liu Y (2017) An integrated AMBBR and IFAS-SBR process for municipal wastewater treatment towards enhanced energy recovery, reduced energy consumption and sludge production. Water Res 110:262–269. https://doi.org/10.1016/j.watres.2016.12.031

Hasan Chowdhury S, Maung Than Oo A (2012) Study on electrical energy and prospective electricity generation from renewable sources in Australia. Renew Sust Energ Rev 16:6879–6887. https://doi.org/10.1016/j.rser.2012.07.015

Hasan MH, Muzammil WK, Mahlia TMI, Jannifar A, Hasanuddin I (2012) A review on the pattern of electricity generation and emission in Indonesia from 1987 to 2009. Renew Sust Energ Rev 16:3206–3219. https://doi.org/10.1016/j.rser.2012.01.075

Huong LQ, Forslund A, Madsen H, Dalsgaard A (2014a) Survival of Salmonella spp. and fecal indicator bacteria in Vietnamese biogas digesters receiving pig slurry. Int J Hyg Environ Health 217:785–795. https://doi.org/10.1016/j.ijheh.2014.04.004

Huong LQ, Madsen H, Anh LX, Ngoc PT, Dalsgaard A (2014b) Hygienic aspects of livestock manure management and biogas systems operated by small-scale pig farmers in Vietnam. Sci Total Environ 470:53–57. https://doi.org/10.1016/j.scitotenv.2013.09.023

Jensen PD, Astals S, Lu Y, Devadas M, Batstone DJ (2014) Anaerobic codigestion of sewage sludge and glycerol, focusing on process kinetics, microbial dynamics and sludge dewaterability. Water Res 67:355–366. https://doi.org/10.1016/j.watres.2014.09.024

Jiang X, Sommer SG, Christensen KV (2011) A review of the biogas industry in China. Energy Policy 39:6073–6081. https://doi.org/10.1016/j.enpol.2011.07.007

Jin X, Li X, Zhao N, Angelidaki I, Zhang Y (2017) Bio-electrolytic sensor for rapid monitoring of volatile fatty acids in anaerobic digestion process. Water Res 111:74–80. https://doi.org/10.1016/j.watres.2016.12.045

Kampanatsanyakorn K, Holasut S, Kachanadul P (2013) Upgrading of biogas to marketable purified methane exploiting microalgae farming. Google Patents

Kavitha S, Yukesh Kannah R, Yeom IT, Do K-U, Banu JR (2015) Combined thermo-chemo-sonic disintegration of waste activated sludge for biogas production. Bioresour Technol 197:383–392. https://doi.org/10.1016/j.biortech.2015.08.131

Kim M-S et al (2016) More value from food waste: lactic acid and biogas recovery. Water Res 96:208–216. https://doi.org/10.1016/j.watres.2016.03.064

Kovács E et al (2015) Augmented biogas production from protein-rich substrates and associated metagenomic changes. Bioresour Technol 178:254–261. https://doi.org/10.1016/j.biortech.2014.08.111

Lantz M (2012) The economic performance of combined heat and power from biogas produced from manure in Sweden – a comparison of different CHP technologies. Appl Energy 98:502–511. https://doi.org/10.1016/j.apenergy.2012.04.015

Lauer M, Thrän D (2017) Biogas plants and surplus generation: cost driver or reducer in the future German electricity system? Energy Policy 109:324–336. https://doi.org/10.1016/j.enpol.2017.07.016

Lijó L, Malamis S, González-García S, Fatone F, Moreira MT, Katsou E (2017) Technical and environmental evaluation of an integrated scheme for the co-treatment of wastewater and domestic organic waste in small communities. Water Res 109:173–185. https://doi.org/10.1016/j.watres.2016.10.057

Lu X, Zhen G, Ni J, Kubota K, Li Y-Y (2017) Sulfidogenesis process to strengthen re-granulation for biodegradation of methanolic wastewater and microorganisms evolution in an UASB reactor. Water Res 108:137–150. https://doi.org/10.1016/j.watres.2016.10.073

Makareviciene V, Sendzikiene E (2015) Technological assumptions for biogas purification. Environ Technol 36:1745–1750. https://doi.org/10.1080/09593330.2015.1008585

Matassa S, Boon N, Verstraete W (2015) Resource recovery from used water: the manufacturing abilities of hydrogen-oxidizing bacteria. Water Res 68:467–478. https://doi.org/10.1016/j.watres.2014.10.028

Mohseni F, Magnusson M, Görling M, Alvfors P (2012) Biogas from renewable electricity – increasing a climate neutral fuel supply. Appl Energy 90:11–16. https://doi.org/10.1016/j.apenergy.2011.07.024

Nghiem LD, Koch K, Bolzonella D, Drewes JE (2017) Full scale co-digestion of wastewater sludge and food waste: bottlenecks and possibilities. Renew Sust Energ Rev 72:354–362. https://doi.org/10.1016/j.rser.2017.01.062

Pham CH, Triolo JM, Sommer SG (2014) Predicting methane production in simple and unheated biogas digesters at low temperatures. Appl Energy 136:1–6. https://doi.org/10.1016/j.apenergy.2014.08.057

Pham CH et al (2017) Biogas production from steer manures in Vietnam: effects of feed supplements and tannin contents. Waste Manage. https://doi.org/10.1016/j.wasman.2017.08.002

Rennuit C, Sommer S (2013) Decision support for the construction of farm-scale biogas digesters in developing countries with cold seasons. Energies 6:5314

Rios M, Kaltschmitt M (2016) Electricity generation potential from biogas produced from organic waste in Mexico. Renew Sust Energ Rev 54:384–395. https://doi.org/10.1016/j.rser.2015.10.033

Roubík H, Mazancová J, Banout J, Verner V (2016) Addressing problems at small-scale biogas plants: a case study from central Vietnam. J Clean Prod 112:2784–2792. https://doi.org/10.1016/j.jclepro.2015.09.114

Roubík H, Mazancová J, Phung LD, Banout J (2017) Current approach to manure management for small-scale Southeast Asian farmers – Using Vietnamese biogas and non-biogas farms as an example. Renew Energy. https://doi.org/10.1016/j.renene.2017.08.068

Santos IFS, Barros RM, Tiago Filho GL (2016) Electricity generation from biogas of anaerobic wastewater treatment plants in Brazil: an assessment of feasibility and potential. J Clean Prod 126:504–514. https://doi.org/10.1016/j.jclepro.2016.03.072

Schoen ME, Xue X, Wood A, Hawkins TR, Garland J, Ashbolt NJ (2017) Cost, energy, global warming, eutrophication and local human health impacts of community water and sanitation service options. Water Res 109:186–195. https://doi.org/10.1016/j.watres.2016.11.044

Shafie SM, Mahlia TMI, Masjuki HH, Ahmad-Yazid A (2012) A review on electricity generation based on biomass residue in Malaysia. Renew Sust Energ Rev 16:5879–5889. https://doi.org/10.1016/j.rser.2012.06.031

Surendra KC, Takara D, Hashimoto AG, Khanal SK (2014) Biogas as a sustainable energy source for developing countries: opportunities and challenges. Renew Sust Energ Rev 31:846–859. https://doi.org/10.1016/j.rser.2013.12.015

Thien Thu CT, Cuong PH, Hang LT, Chao NV, Anh LX, Trach NX, Sommer SG (2012) Manure management practices on biogas and non-biogas pig farms in developing countries – using livestock farms in Vietnam as an example. J Clean Prod 27:64–71. https://doi.org/10.1016/j.jclepro.2012.01.006

Ullah Khan I, Hafiz Dzarfan Othman M, Hashim H, Matsuura T, Ismail AF, Rezaei-DashtArzhandi-M, Wan Azelee I (2017) Biogas as a renewable energy fuel – a review of biogas upgrading, utilisation and storage. Energy Convers Manag 150:277–294. https://doi.org/10.1016/j.enconman.2017.08.035

WorldBank (2013) Vietnam urban wastewater review. © World Bank, Washington, DC

WorldBank (2014) Socialist Republic of Vietnam: review of urban water and wastewater utility reform and regulation. © World Bank, Washington, DC

Index

© Springer International Publishing AG, part of Springer Nature 2018   185
H.-Y. Chan, K. Sopian (eds.), *Renewable Energy in Developing Countries*,
Green Energy and Technology, https://doi.org/10.1007/978-3-319-89809-4

Printed in the United States
By Bookmasters